Numerical Solutions of Singular Integral Equations and Some of Their Applications

Murman Kublashvili

Title: **Numerical Solutions of Singular Integral Equations and Some of Their Applications**

ISBN: 979-8-89248-904-1

Author: Murman Kublashvili

Cover image: www.pixabay.com

Publisher: Generis Publishing
Online orders: www.generis-publishing.com
Contact email: info@generis-publishing.com

Georgian Technical University

Murman Kublashvili

Numerical Solutions of Singular Integral Equations and Some of Their Applications

Tbilisi 2025

The monograph deals with the numerical solutions of singular integral equations of Cauchy type and some of their applications. The author's goal is to develop a method for solving Cauchy-type singular integral equations numerically, which could be used in the solution of problems in hydrodynamics, aerodynamics, theory of elasticity, mathematical physics, etc.

The book may be useful to anyone interested in approximate calculation methods for one-dimensional singular integral equations with Cauchy kernels, and their applications for solving numerically equations that involve these integrals.

Reviewers:

Chairman of the Scientific Council of the Niko Muskhelishvili Institute of Computational Mathematics at the Georgian Technical University, Head of the Shalva Mikeladze Computational Center, Professor, Doctor of Physical and Mathematical Sciences **Hamlet Meladze.**

Head of the Modeling Department at the Niko Muskhelishvili Institute of Computational Mathematics, Georgian Technical University, Professor, Doctor of Physical and Mathematical Sciences **Douglas Ugulava.**

ISBN 978-9941-36-250-7

Dedicated to the bright memory of my scientific advisor,
Professor Jemal Sanikidze.

Table of Contents

Intoduction

Many problems of application type of mechanics (hydrodynamics, elasticity theory, fracture theory, electrodynamics, and others) are reduced naturally to singular integral equations. Besides, plane problems, which often represent good model ones, are reduced to one-dimensional singular integral equations for which full, sufficiently well investigated theory is given in the known monographs [26, 77].

In recent years, singular integral equations are widely used for numerical solution of application problems of various types [1, 2, 7, 14, 16, 17, 18, 31, 35, 37, 42, 61, 77, 78, 79, 82, 83, 85, 86, 92]. Such equations mostly represent convenient ways for solving many important application problems, which had its reflection in monographs of many known authors.

The efficiency of these equations has been discussed in numerous publications, showing their effectiveness in numerical solving of wide class of application problems. Process of their application to calculation problems can be divided into two stages:

The first stage is construction and investigation of approximate calculation schemes for singular integrals.

The second one is construction and foundation of calculation schemes for integrals with singular kernels.

Despite the close interconnection of these two directions, the problems specific to each of them have not yet been fully resolved. It should be noted that in analytical studies of certain problems, solutions have long been reduced to singular integral equations, as a solid theoretical foundation has been developed for them (presented quite comprehensively in the monographs [26, 77]).

The development of numerical methods for solving these equations began relatively earlier than theoretical studies. The well-known article by M. Lavrentiev, published in 1932 [64], served as an impetus for the intensive development of computational methods in this direction, which began in the 1950s. Alongside individual studies, during this period, works were also published that provided more systematic research on the approximate calculation of various types of singular integrals and the numerical solutions of singular integral equations. Among these publications, the works of G. Pikhteev [88, 89] and V. Ivanov [36] should probably be considered among the earliest. The works of S. Belotserkovsky in aerodynamics [7, 8, 9, 10, 11, 12, 13, 14, 15] deserve special mention, as they laid the foundation for a fundamentally new and now well-known method for the numerical solution of singular integral equations—the "discrete singularities method." Later, this method was significantly developed in the works of S. Belotserkovsky and I. Lifanov [16], as well as by their students and followers [17, 18].

Among the works of other authors dedicated to the approximate calculation of singular integrals and the solution of integral equations containing such integrals, it is necessary to highlight the contributions of I. Boikov [20], B. Gabdulkhayev [22], I. Gandel [23-25], B. Musaev [73-76], J. Sanikidze [93-98], M. Sheshko [109], A. Jishkariani [28-30], and their students.

The significance of the works published in various countries is considerable. Among the notable contributions, the works of F.E. Erdogan, G.D. Gupfta [115], S. Krenki [118-119], as well as those by S. Pressdorf and Schmidt [120], are recognized as highly influential.

A wide list of works in the mentioned thematic area is given in a number of articles, including the review works of B. Gabdulkhayev [22].

It is important to note that the trend of development of numerical methods for solving integral equations started for the second type Fredholm equations with 'better' kernel. For such equations, numerical methods were developed: a) high accuracy schemes for narrow-class equations, where the sought functions can be interpolated by special polinomials or by means of partial sums generated by series of eigenfunctions functions b) Based on rectangular type or analogous sufficiently general formulas.

For singular integral equations, challenges arise in constructing numerical algorithms because a singular integral, in the usual sense, is divergent and is understood in a certain specialized sense (e.g., as a Cauchy principal value; see, for instance, [77]). Therefore, in constructing numerical algorithms for such integral equations, additional challenges of a fundamental nature arise.

Unlike Fredholm equations, the development of computational schemes for singular integral equations is characterized by fundamental peculiarities. If for

Fredholm operators involved in the given equations, a more or less good approximation under the solvability conditions of the given equation ensures the foundation of the approximate scheme (see, for example, [38]), for singular integral equations, in general, a good approximation of the singular operator does not guarantee the foundation of the approximate process. On the other hand, numerical solution schemes based on such approximations are generally more practical compared to regularization methods by many parameters.

First, when solving singular equations, additional difficulties often arise, including computational challenges. Moreover, in some cases, depending on the index of the given singular integral equations, it may not be possible to provide an equivalent regularization (see [26, 77]). Moreover, even when an integral equation is not essentially (in the classical sense) singular but can be represented through singular integrals (see [77, 78]), such equations are often more conveniently solved as singular integral equations (see [97, 98]).

The boundary integral equations method, as recognized, represents one of the most effective techniques for investigation and solving various problems of application type. In this direction, alongside individual works, it is worth noting well-known monographs such as [16, 26, 37, 61, 71, 77, 78, 82, 86, 104]. It is noteworthy to mention the translated collection published in 1978, "The Method of Boundary Integral Equations", under the rubric "New in Foreign Science" (edited by A. Yu. Ilshinsky and G. G. Cherniy), in whose introduction fundamental advantages of the boundary equation method compared to other well-known methods are underlined. It is also worth noting works [19, 34, 81], where numerical solutions to many specific problems in continuum mechanics and elasticity theory are discussed, based on singular integral equations.

Analogous (fracture-related) problems are addressed in monographs [16, 42, 77, 78, 82, 84, 92]. Even though the integral equations method represents a powerful and universal tool for solving problems in mechanics, many questions remain unresolved for these problems, both in terms of numerical solutions and the completeness of solution methods. Examples of such problems include calculating the stress state of a body with cracks of arbitrary shape, as well as problems with discontinuous initial conditions, among others.

Also, for such regions which are bounded with (arbitrary) smooth lines there arise difficulties of calculation type, which are mainly conditioned by dependence of the corresponding bounding contour on kernel of the integral equation. This ultimately influences approximation when the initial equation is changed by this or another approximate equation. In this case, Fredholm-type boundary integral equation is meant, to which majority of problems of theory of elasticity can be reduced with the help of potential method. (see for example [77]).

On the other hand, potential methods give us possibility to represent the corresponding integral equations by Cauchy type integrals without reducing them to Fredholm equations (in broadly accepted sense). Thus, this approach allows us to replace the exact equation on the base of singular integral approximation. Quadrature formulas, when used effectively, significantly simplify the calculation processes and it becomes possible to efficiently use differential properties of the given problem (smoothness of the contour and differential properties of the function given on it).

In practical problems of application type (e.g., in cracks), it is often difficult (or sometimes even impossible) to provide exact parametric equations of the line and only its graphical form is known. In such cases according to experiments, singular integral equations method is efficient, since the calculating schemes constructed by us need only knowledge of nod points of the line. At this, during discretization of the exact equation, only the solution sought needs approximation.

The main goal of the given work consists mainly in constructing such calculation schemes for integrals with Cauchy kernel (with open contours), whose application is effective to numerical solving of integral equations thus or otherwise connected with practical problems. Namely, big attention is paid to such important problems of mechanics as cracks, airfoil and, also to Dirichlet original and modified problems of theory of elasticity.

The primary part of this work is organized into six chapters, which address the development of various approximate computational schemes for Cauchy-kernel singular integrals. These schemes are subsequently applied to integral equations that are essential for solving numerous significant problems in mechanics. The work can be conditionally divided into two parts. The first part (Chapters I, II, III) focuses on the development of high-precision numerical algorithms for singular integrals and corresponding integral equations involving open contours, based on appropriate theoretical foundations.

The second part (Chapters IV, V, and VI) addresses the numerical solutions of significant applied problems, such as crack mechanics, fracture mechanics, aerodynamics, hydrodynamics, and fundamental boundary-value problems of elasticity theory. These solutions leverage the algorithms developed in the first part.

The work includes corresponding graphical illustrations and tables, demonstrating the effectiveness of the proposed algorithms compared to other known algorithms.

The first chapter is dedicated to the construction and foundation of various computational processes for singular integrals with open, smooth or piecewise smooth contours, mainly possessing non-smooth density. It should be noted that even for smooth contours, the problem of approximation of singular integrals involves numerous complexities, particularly when such approximation is associated with solving integral equations containing such integrals.

In this study, the construction of various computational processes aligns with specific methods for approximating Cauchy-kernel singular integrals. These methods enable, in many cases, the construction of sufficiently high-precision schemes, even under the assumption that the smoothness of the contour may be violated at certain isolated points along the contour.

Such cases frequently arise in practical problems of applied nature.

In Section 1.1 of the first chapter, the development of approximate computational schemes for singular integrals with open contours having power-type density is discussed.

Relevant error estimate is obtained. Unlike previously known estimations, the obtained estimation enables us to define behavior of the residual term at the endpoints of integration line. By this the problem formulated in the known monograph of academician S. Belotserkovski and I. Lifamov [16] which consists in construction of such high-accuracy calculation scheme for singular integrals with open, arbitrary smooth contour, giving uniform error estimation on the whole contour including the ends, is solved.

In more detail, the following singular integral is considered:

$$S^{(\alpha,\beta)}(\varphi; t_0) = \frac{1}{\pi i} \int_L (t-a)^\alpha (t-b)^\beta \frac{\varphi(t)dt}{t-t_0}, \qquad t_0 \in L,$$

where $L = ab$ is a smooth open contour with ends a and b given on the complex plane by parametric equation $t = t(s)$ $(s_a \leq s \leq s_b)$.

$$\rho(t) = (t - a)^\alpha (t - b)^\beta$$

is a weight multivalued function. Parameters α and β take on a values from the set $\{1/2\,;-1/2\}$. Under $\rho(t)$ we mean one fixed branch of this multivalued function, and φ is an arbitrary function from class $H_\alpha^{(r)}(L)$ $(r \geq 0,\ 1/2 < \alpha \leq 1)$.

Let us consider that n is an arbitrary natural number, and the natural number m is constantly fixed, for approximate value of $S^{(\alpha,\beta)}(\varphi;t_0)$ we take the expression

$$S_n^{(\alpha,\beta)}(\varphi;t_0) = L_v(\varphi;t_0) \cdot \delta^{(\alpha,\beta)}(t_0) + \sum_{\sigma=1}^{n}\sum_{k=1}^{m} \frac{p_{\sigma k}^{(\alpha,\beta)}}{t_0 - t_{\sigma k}}[L_v(\varphi;t_0) - \varphi(t_0)]$$

$(t_0 \in \tau_v\tau_{v+1}; t_0 \neq t_{\sigma k}; v = 1,2,\cdots,n)$,
where

$$\delta^{(\alpha,\beta)}(t_0) = \begin{cases} 0, & \alpha = \beta = -1/2 \\ 1, & \alpha = -\beta = 1/2 \\ t_0, & \alpha = \beta = 1/2 \end{cases}$$

$$\tau_\sigma, t_{\sigma k}\ (\sigma = 1,2,\dots,n;\ k = 1,2,\dots,m)$$

are division points on contour L. $L_v(\varphi;t_0)$ is a Lagrange interpolating polinomial with division points $t_{vk}(v = \overline{1,n};\ k = \overline{1,m})$

$$p_{\sigma k}^{(\alpha,\beta)} = \frac{1}{\pi i}\int_{\tau_\sigma\tau_{\sigma+1}} (t-a)^\alpha(t-b)^\beta \frac{\omega_\sigma(t)dt}{(t-t_{\sigma k})\omega_\sigma'(t_{\sigma k})}, (\omega_\sigma(t)$$
$$= \prod_{k=1}^{m}(t-t_{\sigma k}),(\alpha,\beta = \pm 1/2).$$

At the points $t_0 = t_{\sigma k}$, the sum $S_n^{(\alpha,\beta)}(\varphi;t_0)$ can be written in a more convenient form for computation on a computer, allowing us to avoid division by small values when the value of parameter t_0 is close to the nodes.

In the class of functions $H_\alpha^{(r)}(L)$, the approximation of $S^{(\alpha,\beta)}(\varphi;t_0)$ can be achieved by the corresponding quadrature formula $S_n^{(\alpha,\beta)}(\varphi;t_0)$, at any point $t_0 \in ab$ (with t_0, not coinciding with the end-points a and b), the error is estimated as

$O\left(\dfrac{\ln n}{n^{r+\alpha-1/2}}\right)$ $(1/2<\alpha\leq1, r\geq0)$. However, in some cases, (e.g. crack problems, flow and airfoil problems), such estimation is needed which characterizes more accurately the behavior of the process near the endpoints a and b.

As it follows from the behavior of the singular integral in the neighborhood of the endpoints (see N. Muskhelishvili [77]) for arbitrary $\in H_\alpha(L)$; $\alpha > 1/2$, there exists $\lim\limits_{t_0\to c} S^{(\alpha,\beta)}(\varphi; t_0)$, where c does not coincide with a or b. In this connection the question of uniform approximation of the integral $S^{(\alpha,\beta)}(\varphi; t_0)$ along the whole line L=ab including the endpoints, arises.

For any $t_0 \in ab$ the following estimation is true:

$$\left|S^{(\alpha,\beta)}(\varphi; t_0) - S_n{}^{(\alpha,\beta)}(\varphi; t_0)\right|$$
$$\leq (A_1 + A_2 \ln n)\frac{1}{n^{r+\alpha-1/2}} \quad (n > 1, 1/2 < \alpha \leq 1),$$

where A_1, A_2 are constants, not depending on t_0.

In §1.2 singular integral of type

$$S(\varphi; t_0) = \frac{1}{\pi i}\int_L \frac{\varphi(t)dt}{t - t_0}, \qquad (t_0 \in L) \tag{0.1}$$

is considered where L represents piece-wise smooth contour with $q_1, q_2, \ldots, q_e$ corner points, Approximating sum $S_{n,q_1.q_2,\ldots,q_l}(\varphi; t_0)$ is constructed for such integral:

$$S_{n,q_1.q_2,\ldots,q_l}(\varphi; t_0) =$$
$$= \sum_{\sigma=1}^{n}\left\{\varphi(t_0)\lambda_{r\sigma}(t_0) + \sum_{k=1}^{m}\frac{p_{r\sigma k}}{t_0 - t_{r\sigma k}}[L_{rv}(\varphi; t_0) - \varphi(t_{r\sigma k})]\right.$$
$$+ \sum_{i=1}^{l}\left[\varphi(q_i)\lambda_{i\sigma}(t_0) + \sum_{k=1}^{m}\frac{\Omega_{i\sigma k}(t_0; q_i)}{q_i - t_{i\sigma k}}\right] \times [L_{i,0}(\varphi; q_i) - \varphi(t_{i\sigma k})]$$
$$+ \varphi(q_r)\lambda_{r-1\sigma}(t_0)$$
$$+ \sum_{k=1}^{m}\frac{\Omega_{r-1\sigma k}(t_0, q_r)}{q_r - t_{r-1\sigma k}}[L_{r-1,n-1}(\varphi; q_r) - \varphi(t_{r-1\sigma k})]\right\},$$

where

$$\Omega_{i\sigma k}(t_0, q_i) = p_{i\sigma k} + \frac{t_0 - q_i}{(t_0 - t_{i\sigma k})\omega'_{i\sigma}(t_{i\sigma k})}[\omega_{i\sigma}(t_0)[\gamma_{i\sigma}(t_0) + \lambda_{i\sigma}(t_0)]$$
$$- p_{i\sigma k}\omega'_{i\sigma}(t_{i\sigma k})],$$

$$\lambda_{i\sigma}(t_0) \equiv \frac{1}{\pi i} \ln\frac{\tau_{i\sigma+1} - t_0}{\tau_{i\sigma} - t_0}; \gamma_{i\sigma}(t_0) \equiv \sum_{j=1}^{m} \frac{p_{i\sigma j}}{t_0 - t_{i\sigma j}}; p_{i\sigma k}$$

$$= \frac{1}{\pi i}\int_{\tau_{i\sigma}\tau_{i\sigma+1}} \frac{\omega_{i\sigma}(t)dt}{(t - t_{i\sigma k})\omega'_{i\sigma}(t_{i\sigma k})},$$

$$(i = 1,2,3, \dots, r - 1, r + 1, \dots, l) \quad (t_0 \in q_r q_{r+1}),$$

Here $t_{i\sigma k}(i = 1,2, \dots, l; \ \sigma = 1,2, \dots, n; \ k = 1,2, \dots, m)$ are division points on the arc $q_i q_{i+1} \subset L$ (if L is closed, under $q_r q_{r+1}$ we mean the least arc with ends q_r and q_{r+1}.

While constructing the corresponding approximation formulas, in addition to achieving a high degree of accuracy, one of the important requirements is that they should be effectively applied to the numerical solution of practical problems.

In §1.3 a singular integral

$$S^{(\vec{\alpha},\vec{\beta})}_{q_1,q_2,\dots,q_p}(\varphi; t_0) = S^{(\alpha_i,\beta_i)}(\varphi, t_0), \quad t_0 \in q_i q_{i+1}$$
$$S^{(\alpha_i,\beta_i)}(\varphi; t_0) = \frac{1}{\pi i}\int_{q_i q_{i+1}} (t - q_i)^{\alpha_i}(t - q_{i+1})^{\beta_i}\frac{\varphi_i(t)dt}{t - t_0}$$
$$(i = 1,2, \dots, p - 1),$$

$\vec{\alpha} = (\alpha_1, \alpha_2, \dots, \alpha_p), \vec{\beta} = (\beta_1, \beta_2, \dots, \beta_p)$ is considered.

Under $S^{(\vec{\alpha},\vec{\beta})}_{q_1,q_2,\dots,q_p}(\varphi; q_i)$ we always mean that

$$S^{(\vec{\alpha},\vec{\beta})}_{q_1,q_2,\dots,q_p}(\varphi; q_i) = S^{(\vec{\alpha},\vec{\beta})}_{q_1,q_2,\dots,q_p}(\varphi; q_i^+)$$
$$(i = 1,2, \dots, p - 1).$$

Under the weight functions $\rho_i(t) = (t - q_i)^{\alpha_i}(t - q_{i+1})^{\beta_i}$ $(i = 1,2, \dots, l)$ we mean some fixed branch of the corresponding function.

Such integrals are seen in plane problems of theory of elasticity and cracks for non-isotropic bodies. The following quadrature formula

$$S_{n,q_1,q_2,\ldots,q_p}^{(\vec{\alpha},\vec{\beta})}(\varphi; t_0) = S_n^{(\alpha_i,\beta_i)}(\varphi; t_0), \quad t_0 \in q_i q_{i+1}$$

$$(i = 1,2,\ldots,p-1),$$

is constructed for them where the approximating algorithm $S_n^{(\alpha_i,\beta_i)}(\varphi; t_0)$ for the corresponding arcs $q_i q_{i+1}$ is defined according to §1.1.

Parameters α_i, β_i, as mentioned above, take one of the values from the set $\{\pm 1/2\}$.

It is proven that when $\varphi_i \in H_\alpha^{(r)}(q_i, q_{i+1})$ $(i = 1,2,\ldots,p)$, for arbitrary $t_0 \in L$ the estimation

$$\left| S_{q_1,q_2,\ldots,q_p}^{(\vec{\alpha},\vec{\beta})}(\varphi; t_0) - S_{n,q_1,q_2,\ldots,q_p}^{(\vec{\alpha},\vec{\beta})}(\varphi; t_0) \right| \leq \frac{A_r \ln n}{n^{r+\alpha-1/2}}, \quad \left(\frac{1}{2} < \alpha \leq 1\right)$$

is valid where A_r is a constant independent of parameter t_0.

The combination of the results from §§1.1–1.3 allows us to develop high-accuracy schemes for numerical solutions of applied problems (Chapters IV–VI).

Although the sums $S_n^{(\alpha,\beta)}(\varphi; t_0)$, $S_{n,q_1,q_2,\ldots,q_n}(\varphi; t_0)$, $S_{n,q_1,q_2,\ldots,q_p}^{(\vec{\alpha},\vec{\beta})}(\varphi; t_0)$ realize sufficiently high accuracy approximation of the corresponding integrals, their justification for solution of the singular integral equations is rather difficult, and in this connection, in §1.4, the sums $L_n\left[S_n^{(\alpha,\beta)}\varphi; t_0\right]$, $L_n\left[S_{n,q_1,q_2,\ldots,q_n}\varphi; t_0\right]$ and $L_n\left[S_{n,q_1,q_2,\ldots,q_p}^{(\vec{\alpha},\vec{\beta})}(\varphi; t_0)\right]$ with different structure are constructed.

It is demonstrated that these sums converge to the corresponding integral in a relatively stronger sense. Specifically, in §1.4, it is shown that under the conditions outlined above for φ, the following estimates hold true

$$\left\| S^{(\alpha,\beta)}(\varphi; t_0) - L_n\left[S_n^{(\alpha,\beta)}\varphi; t_0\right] \right\|_{H_\beta} \leq \frac{A_r \ln n}{n^{r+\alpha-\beta-1/2}}$$

$$\left\| S(\varphi; t_0) - L_n\left[S_{n,q_1,q_2,\ldots,q_n}\varphi; t_0\right] \right\|_{H_\beta} \leq \frac{A_r' \ln n}{n^{r+\alpha-\beta}} \tag{0.2}$$

$$\left\|S_{q_1,q_2,\dots,q_p}^{(\vec{\alpha},\vec{\beta})}(\varphi;t_0) - L_n\left[S_{n,q_1,q_2,\dots,q_p}^{(\vec{\alpha},\vec{\beta})}(\varphi;t_0)\right]\right\|_{H_\beta} \leq \frac{A'' \ln n}{n^{r+\alpha-\beta-1/2}},$$

$$(n > 1; \quad r \geq 0; \quad \beta < \alpha - 1/2),$$

where $\|\cdot\|_{H_\beta}$ denotes the norm with the exponent β in the metric of the Hölder function space, and $A'_r, A''_r, A_z,$ are constants that depend solely on the Hölder exponent of the function $\varphi^{(r)}$.

§1.5 is dedicated to constructing a specific approximation scheme of the type of the discrete singularity method for the singular integral $S^{(\alpha,\beta)}(\varphi;t_0)$, discussed in §1.1, and to estimating the error in the case of open smooth contours.

The question of construction of modified schemes of discrete singularities type when L is a closed Liapunov line is described in J. Sanikidze's work [99]. For further development of the appropriate approach, the subsequent article by the same author [100] presents an analogous exact variant of these schemes, providing an error estimate for the certain smooth classes of density functions ϕ and for one type of singular integral equations a process of numerical solution is founded.

It should be noted that the accuracy order, which is achieved in the approximation process, by the method described, is determined by the differential properties of the density function $\varphi(t)$, regardless other structural parameters (except the fact that L-contour is of Liapunov type) and also by accuracy order of the constructed approximate formula itself.

For any natural number n, consider the interval $[s_a, s_b]$. We divide it into 2n equal parts, (n>2). On L=ab, define the system of points $\{\tau_p\}_{p=1}^{2n}$ which divides the segment L into 2n equal (length) intervals.

In the singular integral $S^{(\alpha,\beta)}(\varphi;t_0)$, change the function $[\varphi(t) - \varphi(t_0)](t - t_0)^{-1}$ at points $t = \tau_{2\sigma-1}, t = \tau_{2\sigma+1}$ on each arc $[\tau_{2\sigma-1}, \tau_{2\sigma+1}]$ by Lagrange linear interpolation formula.

Based on this we can write:

$$S^{(\alpha,\beta)}(\varphi;t_0) \approx D_n^{(\alpha,\beta)}(\varphi;t_0)$$
$$= \delta^{(\alpha,\beta)} \cdot \varphi(t_0) + q_1{}^{(\alpha,\beta)}\chi(\tau_1,t_0) + p_{2n-1}^{(\alpha,\beta)}\chi(\tau_{2n-1},t_0) +$$

$$+ \sum_{\sigma=1}^{n-1} \left(p_{2\sigma-1}^{(\alpha,\beta)} + q_{2\sigma+1}^{(\alpha,\beta)} \right) \cdot \chi(\tau_{2\sigma+1}, t_0) \qquad (0.3)$$

where

$$\chi(\tau_\sigma, t_0) = \frac{\varphi(\tau_\sigma) - \varphi(t_0)}{\tau_\sigma - t_0},$$

$$p_{2\sigma-1}^{(\alpha,\beta)} = \frac{1}{\pi i} \int_{\tau_{2\sigma-1}\tau_{2\sigma+1}} (t-a)^\alpha (t-b)^\beta \frac{t - \tau_{2\sigma-1}}{\tau_{2\sigma+1} - \tau_{2\sigma-1}}\, dt\,;$$

$$q_{2\sigma+1}^{(\alpha,\beta)} = \frac{1}{\pi i} \int_{\tau_{2\sigma-1}\tau_{2\sigma+1}} (t-a)^\alpha (t-b)^\beta \frac{t - \tau_{2\sigma-1}}{\tau_{2\sigma+1} - \tau_{2\sigma-1}}\, dt,$$

$$(\sigma = 1,2,\cdots,n).$$

Parameters α and β, as earlier, take on one of values from the set $\{\pm 1/2\}$.

Next, based on the special expansion of the density $\varphi(t)$, the integral representation of the residual term is given, from which it specifically follows that if $\varphi(t)$ has a second-order bounded derivative, the given quadratic formula provides an approximation order $O\left(n^{-3/2+\varepsilon}\right)$, where ε is an arbitrarily small positive number.

Numerical examples are given well demonstrating high accuracy of algorithms constructed by us.

In the second chapter of this work, the application of approximation formulas constructed in the previous chapter to numerical solution of singular integral equation of a specific class is presented. Many important applied problems are reduced to such equations, and the Dirichlet modified problem among them, to which in its turn the problem of finding conformal mapping function is reduced, as well as many other practical (cracks, torsion, flow and airfoil) problems.

Namely, in § 2.1

$$\frac{1}{\pi i} \int_L \frac{\varphi_0(t)dt}{t - t_0} + \frac{1}{\pi i} \int_L K(t_0, t)\varphi_0(t)dt = f(t_0) \qquad (0.4)$$

is investigated.

The question of numerical solution of the first type singular integral equation with open contour where $f(t)$, $K(t_0, t)$ are functions from Hölder class defined on L=ab.

In the given section, the discussed scheme is utilized for the numerical solution of first-order singular integral equation with arbitrary open contours, under the assumption that the exact equation in the considered class has a unique solution.

Under specific assumptions regarding the right-hand side and the kernel, it is possible to achieve a sufficiently high order of convergence.

The construction of the corresponding computational algorithm is based on the approximation of singular integrals discussed in §1.1.

Based on the result obtained in §1.4, which establishes the convergence of quadrature processes for singular integrals with open contours in H_β metric, it is possible to validate the coresponding computational processes.

At the same time, the constructed quadrature processes allow us to relatively easily develop corresponding practical computational schemes.

In addition, these schemes have the important property that the error in computing the kernel K and the right-hand side f, as n→∞, practically grows very slowly.

In §2.2, the numerical solution of equation (0.4) is analyzed in cases where its solution is not uniquely determined or where solving the equation requires additional conditions for the right-hand side f. For such equations, high-accuracy schemes are constructed.

In §2.3, various numerical schemes for solving equations with discrete singularities particularly focusing on the first-order singular integrals are discussed. Based on algorithms developed in §1.5, estimations of difference between solutions of exact and the corresponding approximate equations is discussed.

In §2.4 approximate processes are constructed for the first type full singular integral equation.

In the third chapter, approximate computational schemes are constructed for Cauchy type integrals and their derivatives in the case of open contours. In the work of J. Sanikidze and K. Ninidze [102], approximation schemes have been developed for solving Cauchy-type integrals and their derivatives for closed Lapunov contours.

In the preceding study, similar schemes have been developed for weighted Cauchy-type integrals in the case of open contours. The mentioned schemes are based on a specific process of density approximation and ensure uniform error estimates across the entire domain, including the boundary.

Using these formulas, it is possible to determine stresses and displacements in problems of elasticity theory for arbitrary, including boundary-adjacent, points. This is achieved without loss of accuracy.

This aspect is significant in solving numerical crack problems. In such problems, it is crucial to determine stress intensities in the vicinity of crack tips. Specifically, this applies when the Cauchy-type integral involves a z-point that approaches the endpoints of the integration contour. As is well known, in crack-related problems, it is essential to evaluate stresses in small proximity to the crack tips.

Thus, the integral

$$F(\rho; \varphi; z) = \frac{1}{2\pi} \int_L \rho(t) \frac{\varphi(t)}{t - z} dt,$$

is studied, where $L = ab$ is an open smooth contour, $\rho(t) = (t - a)^\alpha (t - b)^\beta$, $\alpha, \beta \in \{-1/2; +1/2\}$ is a weight function with a fixed branch meant.

For such integral, the main source of construction is special approximation of the density function $\varphi(t)$ namely, its approximation is realized by the following formula

$$\varphi(t) \approx L_V(\varphi; t_0) + (t - t_0) \sum_{k=0}^{1} l_{vk}(t) \frac{\varphi(\tau_{\sigma+k}) - L_{\sigma k}(\varphi; t_0)}{\tau_{\sigma+k} - t_0},$$

where $L_v(\varphi; t_0) = l_{v0}(t_0)\varphi(\tau_v) + l_{v1}(t_0)\varphi(\tau_{v+1})$ is the Lagrange interpolation formula, $L_{vk}(\varphi; t_0)$ denotes

$$L_{\sigma k}(\varphi; t) = \begin{cases} \varphi(t_0), & a + k \neq v, \quad v + 1; \\ L_v(\varphi; t_0), & \sigma + k = v, \quad v + 1, \quad (t_0 \in \tau_v \tau_{v+1}) \end{cases}$$

Here, t_0 is till considered as any point on the L-contour, and it is later somehow connected to the choice of point z. Based on the given approximation of the density $\varphi(t)$ and similar approximations, quadrature formulas are derived for both the integral $F(\rho; \varphi; z)$ and its derivatives $F'(\rho; \varphi; z), F''(\rho; \varphi; z)$ (with values of density $\varphi(t)$). The necessity of calculating these derivatives arises in the practical solution of the problem, as we have previously discussed.

Namely, the quadrature formula for calculation of $F(\rho; \varphi; z)$ has the following form:

$$F(\rho; \varphi; z) \approx F(\rho; 1; z) L_v(\varphi; t_0) + p_{v-10}(\rho; t_0; z) \frac{\varphi(\tau_{v-1}) - \varphi(t_0)}{\tau_{v-1} - t_0} \times$$

$$\times \left[p_{v-11}(\rho; t_0; z) + p_{v0}(t_0; z) + p_{v1}(\rho; t_0; z) \right.$$

$$\left. + p_{v+10}(\rho; t_0; z) \frac{\varphi(\tau_{v+1}) - \varphi(\tau_v)}{\tau_{v+1} - \tau_v} \right] +$$

$$+ p_{v+11}(\rho; t_0; z) \frac{\varphi(\tau_{v+2}) - \varphi(t_0)}{\tau_{v+2} - t_0} + \sum_{\sigma \neq v \pm 1}^{n} \sum_{k=0}^{1} p_{\sigma k}(\rho; t_0; z) \frac{\varphi(\tau_{\sigma+k}) - \varphi(t_0)}{\tau_{\sigma+k} - t_0},$$

At this, it is clear that the right-hand side is already explicitly defined for values $t_0 = \tau_v$, $t_0 = \tau_{v+1}$ as well. Here

$$p_{\sigma k}(\rho; t_0; z) = \frac{1}{2\pi i} P_\sigma(\rho) + \frac{z - t_0}{2\pi i} \int_{\tau_\sigma \tau_{\sigma+1}} \rho(t) \frac{l_{\sigma k}(t) dt}{t - z},$$

$$P_\sigma(\rho) = \frac{1}{2\pi i} \int_{\tau_\sigma \tau_{\sigma+1}} \rho(t) dt$$

$$(\sigma = 1, 2, \cdots, n).$$

The following scheme is used for approximation the derivative $F'(\rho; \varphi; z)$

$$\varphi(t) \approx \varphi(t_0) + (t - t_0)\varphi(t_0, t_1)(t - t_1) \sum_{k=0}^{1} l_{\sigma k} \frac{\varphi(\tau_{\sigma+k}, t_1)}{\tau_{\sigma+k} - t_0},$$

where like t_0, $t_1 \in L \equiv ab$ ($t_1 \neq a, b$) is yet taken arbitrarily, $\varphi(t_0, t_1)$, $\varphi(\tau_{\sigma+k}, t_1)$ are first order divided differences. As in the previous case, from the given approximation we get quadrature formula for derivative of the integral

$$F'(\rho; \varphi; z) \approx F'(\rho; 1; z)L_v(\varphi; t_0) + F'(\rho; t - t_0; z)\frac{\varphi(\tau_{v+1}) - \varphi(\tau_v)}{\tau_{v+1} - \tau_v} +$$

$$+ \frac{p_{v-10}(\rho; t_0; t_1; z)}{\tau_{v-1} - t_0}\left\{\frac{\varphi(\tau_{v-1}) - \varphi(t_1)}{\tau_{v-1} - t_1} - \frac{\varphi(\tau_{v+1}) - \varphi(\tau_v)}{\tau_{v+1} - \tau_v}\right\} +$$

$$+ \frac{p_{v+11}(\rho, t_0, t_1, z)}{\tau_{v+2} - t_0}\left\{\frac{\varphi(\tau_{v+2}) - \varphi(t_1)}{\tau_{v+2} - t_1} - \frac{\varphi(\tau_{v+1}) - \varphi(\tau_v)}{\tau_{v+1} - \tau_v}\right\} +$$

$$+ \sum_{\sigma \neq v, v\pm 1;}^{n} \sum_{\kappa=0}^{1} \frac{p_{\sigma\kappa}(\rho, t_0, t_1, z)}{\tau_{v+2} - t_0} \times \left\{\frac{\varphi(\tau_{v+\kappa}) - \varphi(t_1)}{\tau_{v+\kappa} - t_1} - \frac{\varphi(\tau_{v+1}) - \varphi(\tau_v)}{\tau_{v+1} - \tau_v}\right\}$$

In the last expression $t_0, t_1 \in \tau_v \tau_{v+1}$ (at this we can also suppose their equality to division points), $t_0 = t_1$ equality is also possible. According to this, coefficients $p_{\sigma\kappa}(\rho, t_0, t_1, z)$ will take the following form

$$p_{\sigma\kappa}(\rho, t_0, t_1, z) = \frac{1}{2\pi\iota}\int_{\tau_\sigma \tau_{\sigma+1}} \rho(t)\frac{(t - t_0)^2 l_{\sigma\kappa}(t)dt}{(t - z)^2}.$$

The obtained integrals can be easily calculated.

In the following paragraph, (3.2), an error estimation is derived, which arises due to replacing the exact values of $F(\rho; \varphi; z)$ and its derivatives with their approximate values.

It has been established that if, for a given z, the parameter t_0 is chosen so that the distance between z and t_0 does not exceed the minimal distance from z to the boundary points, then on the class of functions $\varphi(t)$ with a bounded second derivative on L, a convergence rate of $O(h^{3/2}\ln n)$ is ensured.

It should be noted that if we take the limit as $z \to L$, the discussed formula provides a quadrature formula for the singular integral with kernel $(t - t_0)^{-1}$. This enables us to compute the Cauchy-type integral and its boundary values.

As we can see, to determine the error order of the quadrature formula for $F'(\rho; \varphi; z)$, an approximation is used that employs two free parameters, t_0, t_1. The appropriate selection of these parameters provides an error order of $O(h^{1/2}\ln n)$.

Moreover, this estimate remains valid even in the case when $t_0 \to t_1$. Thus, in the computational formulas, $t_0 = t_1$ is assumed.

A similar approach can be applied to approximate calculation of $F''(\rho; \varphi; z)$ if instead of using two-point linear interpolation on the arcs $\tau_v \tau_{v+1}$ we use three-point quadratic interpolations

Chapter 4 examines fracture problems in brittle bodies with defects using the method of singular integral equations. It determines how stresses should be distributed within bodies during the formation of cracks and holes.

Paragraph (4.1) explores the stressed and deformed state of an elastic plane with rectilinear crack. The stress distribution and displacement functions are represented in the form of Cauchy-type integrals. The computational formulas for intensity coefficients are presented in integral form in the most general case.

The limiting equilibrium state of brittle bodies containing crack-type defects is investigated. The critical value of external loading is determined, at which cracks begin to propagate, i.e. local or complete failure of the body begins. Formulas are provided for calculating the initial crack propagation angle and the critical stress values. Diagrams illustrating the limiting stress are also presented.

§ 4.2 discusses the numerical solution of the problem of a thermally insulated crack. An infinite body is considered, on the surface of which, along a finite strip, a thermally insulated crack is present. The temperature variation along the crack front is known, and it is necessary to determine the temperature change at any point of the body surface.

The mentioned problem is reduced to the first-kind singular integral equation. The temperature distribution on the surface is calculated by weighted Cauchy-type integral, which is implemented with the algorithm developed in Chapter III. The corresponding numerical calculation tables are provided. The calculations show that for specific points (in the very small neighborhood of the ctack endpoints), using the mentioned approximate algorithms, sufficiently high accuracy is achieved.

In paragraph 4.3, the numerical solution of the problem of the out-of-plane shear of the crack on an elastic body is presented.

An infinite solid is investigated under out-of-plane shear, A rectilinear crack of finite length exists on the surface of the body, a self-regulating loading acts on the surface of the crack, and the tension is zero at infinity. In this case also, the stress distribution at any point on the surface of the solid is to be calculated. It is given by weighted Cauchy-type integrals, and the algorithm for their approximate calculation is given in chapter III. The tables are given when the loading at the crack's edge is given by concrete functions.

In Section 4.4, the crack problem is considered when on an infinite plane two equal and symmetric cracks are present, and a loading uder condition of symmetry is given. The mentioned problem is reduced to the first kind singular integral equation, which is approximately solved by approximation schemes constructed in chapter II. Corresponding tables are given.

In Section 4.5, the numerical solution for calculation of a crack perpendicular to the boundary of the semi-plane is considered.

We consider a finite crack in an elastic semi-plane which is perpendicular to the plane boundary and the self-regulating loading is given along the crack edge. This problem is reduced to the singular integral equation of the following type:

$$\frac{1}{\pi}\int_{-1}^{+1}\frac{\varphi_0(\tau)d\tau}{\tau-y}+\frac{1}{\pi}\int_{-1}^{+1}K(\tau,y)\varphi_0(\tau)d\tau=f(y),|y|<1,\qquad(0.5)$$

where

$$K(\tau,y)=\frac{y^2+6y+4\tau y+2\tau-\tau^2+4}{(\tau+y+2)^3}$$

And $f(y)$ is the loading given on the crack sides.

Besides singularity of the kernel $K(\tau,y)$, it has also fixed singularity at the point $\tau=y=-1$. For numerical solution of equation (0.5), a justifiable algorithm in ceitrain functional space is constructed. Corresponding tables for computational results are provided.

Here, a crack problem is discussed, with a point load applied at a fixed point on the crack boundary.

The mentioned problem is reduced to an equation of type (0.5), therefore, for its numerical solution it is natural to use the same algorithm as in the previous case.

In section §4.6, several important crack problems are discussed, which are resuced to the first-order singular integral equations.

The fifth chapter of the work discusses numerical solutions to problems of planar aerodynamics.

In § 5.1, the stationary and non-stationary circulation problems of a wing are analyzed.

In § 5.2, the equations of airfoil are discussed, focusing on the derivation of these equations to such first-kind singular integral equations, in which the integration line (airfoil line) represents an open smooth contour given in parametric form. The characteristics of the line equation are present in the kernel of the equations.

In § 5.3, the numerical solution of the thin airfoil problem is studied given circulatory flow.

For numerical solution of this problem, the scheme of approximate solution of the first kind singular integral equation, constructed in chapter II is offered. Tables of the corresponding numerical experiments are present.

In § 5.4, the same problem is solved using a simplified algorithm, to be more precise, employing discrete singularities method with an increased accuracy.

In §6.1, the numerical solution of the modified Dirichlet problem is discussed, when the crack is given as a smooth open curve of arbitrary shape on the plane.

The mentioned problem is reduced to the first-kind singular integral equation, which is solved by the algorithm developed in Chapter II. This algorithm allows to find values of solution $\{\varphi_j\}_{j=1}^{n}$ at discrete points.

In the continuation of the solution process, interpolation function $\psi_n(t)$ is constructed and in the role of approximate solution we take

$$U_n(x,y) = Re\frac{1}{\pi}\int_{ab} \rho(t)\frac{\psi_n(t)}{t-z}dt \ \ (z \in D),$$

The obtained integral is calculated by formula from §3.1.3.

In the same paragraph, numerical solutions of the mentioned problem are considered in concrete cases, both for the case with parametric open contour and for an arbitrary contour, obtained by computer-based graphical interfaces. The last case may turn out to be of special interest in problems connected with practical applications, numerical results are given in proper tables and are illustrated by corresponding pictures.

Consideration of the cases when the boundary is given not by exact equation, but graphically (e.g., using coordinates of the contour points), shows efficiency of the singular integral equations method, since the numerical schemes constructed in this way need only knowledge of the devision points.

Construction of open smooth contours amd their application to solution of concrete problems given in the work for further construction of mathematical software is done by application package „Spline Toolbox" in „MATLAB" invironment.

From the presented tables it appears that while several approximate processes were sequentially applied (such as solving integral equations through numerical methods, followed by constructing interpolation formula, and finally computing Cauchy integral), the rounding error does not essentially influence on convergent process. This fact indicates once more that the algorithms constructed by us are stable.

In §6.2, the basic boundary problems of the theory of elasticity on plane with finite number of rectilinear cracks, are considered. The components of stresses and displecements of these problems are written in the form of Cauchy-type integrals, whose approximate calculation with high accuracy is possible using algorithms from chapter III.

In §6.3 a numerical solution of a problem of binding of a homogeneous prismatic beam with cracks is considered. This problem is also reduced to the Dirichlet modified problem.

Using the solution to this problem, complex function of the corresponding binding can be represented in the form of the Cauchy-type integral, whose approximate calculation can be realized by algorithm from chapter III.

Chapter I. Some Methods of Approximate Solution of Singular Integrals

§1.1 On approximate calculation of singular integrals with power-type singularities

Consider the following singular integral

$$S(\varphi; t_0, \eta) \equiv \frac{1}{\pi i} \int_L \frac{\varphi(t)dt}{(t-c)^\eta (t-t_0)} \quad (t_0 \in L), \tag{1.1.1}$$

where $L \equiv ab$ is an open smooth contour with endpoints a and b a, given by equation $t = t(s)$, $(s_a \leq s \leq s_b)$, at this, the positive direction is taken along increasing direction of the parameter s. c is either a or b, $\eta = \lambda + i\beta$, $(0 \leq \lambda < 1)$, under $(t-c)^\lambda$ an arbitrary fixed branch of this multivalued function is meant.

It is known that (see [77], §22), for $\eta = \lambda + i\beta \neq 0$ then

$$S(\varphi; t_0, \eta) = \pm \frac{ctg\eta\pi}{2i} \frac{\varphi(c)}{(t_0 - c)^\eta} + \phi^*(t_0),$$

At this, if $\lambda = 0$, then $\phi^*(t_0)$ belongs to the class H (satisfies the Hölder condition) in the neighbourhood of the point c. When $\lambda > 0$, then

$$\phi^*(t_0) = \frac{\phi^{**}(t_0)}{|t_0 - c|^{\lambda_0}}, \quad 0 < \lambda_0 < \lambda,$$

Where $\phi^{**}(t_0)$ belongs to H in the neighbourhood of the point c, the upper sign is taken for $c = a$, and the lower sign for $c = b$. From this reasoning it follows that for $Re\,\eta \neq 0$, $S(\varphi; t_0, \eta)$, in the neighborhoods of the points a and b, may be bounded if and ony if when $\eta = 1/2 + i\beta$, (β is an arbitrary real number). This case is interesting also due to the fact that singular integrals with weights, having square root-type singularities, often occur in applications.

Inversion formulas of singular integrals with open-contours are reduced to such integrals. And by it, as it is known from the general theory of singular integral equations (see [77], ch. V), the structure of solutions of the general first kind singular integral equations at the endpoints of the integration line is apriori determined via singularities of the mentioned type.

This class of equations is particularly significant because many important practical problems are reduced to them (see [77-78], [82], [16]).

Denote

$$S^{(v;\delta)}(\varphi; t_0) = \frac{1}{\pi i} \int_{ab} (t-a)^v (t-b)^\delta \frac{\varphi(t)dt}{t-t_0}, \quad (v, \delta \in \{\pm 1/2\})$$

For the sake of concreteness and simplicity, here, in this § we consider questions of approximation of singular integrals of the following type:

$$S^{(1/2;-1/2)}(\varphi; t_0) \equiv \frac{1}{\pi i} \int_{ab} \sqrt{\frac{t-a}{t-b}} \frac{\varphi(t)dt}{t-t_0}, \qquad (t_0 \in ab, t_0 \neq a, b) \quad (1.1.2)$$

In the quadratic formulas provided below, for approximation of singular integrals, the system of nodes is constructed like it is done for constructing complicated quadratic formulas for a bounded integrable function.

Let n be an arbitrary natural number, divide the interval $[a, b]$ into n equal parts by points

$$s_\sigma = a + \frac{b-a}{n}(\sigma - 1), \quad \sigma = 1, 2, \ldots n + 1.$$

Then, for the given fixed natural number m, divede each segment $[s_\sigma, s_{\sigma+1}]$ by points

$$s_{\sigma k} = s_\sigma + h \cdot x_k \quad \left(h = \frac{b-a}{n}\right), \qquad k = 1, 2, \ldots m,$$

where $\{x_k\}_{k=1}^m$ is a system of given points from the segment $[0,1]$.

Below for definiteness, we mean that the values $x_1, x_2, x_3, \ldots, x_m$ are numerated in ascending order.

Denote $\tau_\sigma = t(s_\sigma)$, $t_{\sigma k} = t(s_{\sigma k})$ $(\sigma = 1, 2, \ldots, n; \ k = 1, 2, \ldots m)$.

Let's mean that $,\nu$ are arbitrary values from $1, 2, \ldots, n$ and $t_0 t_0$ differs from the division points $t_{\sigma k}$, consider the expression

$$\Psi_{\sigma\nu}(\varphi; t, t_0)$$

$$= \varphi(t_0) + \sum_{k=1}^{m} l_{\sigma k}(t) \frac{t - t_0}{t_{\sigma k} - t_0} \cdot \varphi(t_{\sigma k})$$

$$+ \sum_{k=1}^{m} l_{\sigma k}(t) \frac{t - t_0}{t_0 - t_{\sigma k}} \cdot L_\nu(\varphi; t_0),$$

where

$$L_\nu(\varphi; t_0) = \sum_{k_0=1}^{m} l_{\nu k_0}(t_0)\varphi(t_{\nu k_0}), \qquad l_{\sigma k}(t) = \frac{\omega_\sigma(t)}{(t - t_{\sigma k})\omega'_\sigma(t_{\sigma k})},$$

$$\omega_\sigma(t) = \prod_{k=1}^{m}(t - t_{\sigma k})$$

and φ, is a function given on L from Hölder class ($\varphi \in H$).

Let's note that thoroughly definite sense may be given to the expression $\Psi_{\sigma\nu}(\varphi; t, t_0)$ when $t_0 = t_{\sigma k}$ (in this case we must mean that $t_0 \to t_{\sigma k}$).

Indeed, let us assume that the natural number ν is always chosen so, that for the given t_0 the relation $t_0 \in \tau_\nu, \tau_{\nu+1}$ holds. We observe that $t_0 - t_{\sigma k_1} = 0$ ($1 \leq k_1 \leq m$) is possible only when $\sigma = \nu, \nu \pm 1$ (and if simultaneously $x_1 > 0, x_m < 1$ when $\sigma = \nu$).

It is easy to show that e.g. for $\sigma = \nu$ the corresponding expression

$$L_\nu(\varphi; t_0) - \varphi(t_{\nu k_1})$$

in $\Psi_{\nu\nu}(\varphi; t, t_0)$ is divisible by $t_0 - t_{\nu k_1}$ $(t_0 \neq t_{\nu k_1})$. The same occurs also when $\sigma = \nu - 1, \sigma = \nu + 1$ (if, respectively, $x_1 = 0, x_m = 1$).

Now, denote

$$\Phi_n(\varphi, t, t_0) = \psi_{\sigma\nu}(\varphi; t, t_0), t \in \tau_\sigma \tau_{\sigma+1}, t_0 \in \tau_\nu \tau_{\nu+1}$$

$$\sigma = 1, 2, \ldots n; \quad \nu = 1, 2, \ldots n.$$

Based on the above remark, the expression $\Phi_n(\varphi; t, t_0)$ is defined for any value of $t, t_0 \in L$.

Furthermore, the following structural representation of $\Phi_n(\varphi; t, t_0)$ is evident:

$$\Phi_n(\varphi; t, t_0) = \varphi(t_0) + (t - t_0)H_n(\varphi; t, t_0),$$

where

$$H_n(\varphi; t, t_0) = \sum_{k=1}^{m} l_{\sigma k}(t) \cdot \frac{L_\nu(\varphi; t_0) - \varphi(t_{\sigma k})}{t_0 - t_{\sigma k}} \ , \ t \in \tau_\sigma \tau_{\sigma+1}$$

$$t_0 \in \tau_\nu \tau_{\nu+1}, \quad \sigma, \nu = 1,2,3,\dots,n$$

$$H_n\big(\varphi; t, t_{\sigma k_1}\big) = \lim_{t_0 \to t_{\sigma k_1}} H_n(\varphi; t, t_0)$$

for any k_1 $(1 \le k_1 \le m)$.

Change the singular integral $S^{(1/2;-1/2)}(\varphi; t_0)$ by $S_n^{(1/2;-1/2)}(\varphi; t_0)$

$$S_n^{(1/2;-1/2)}(\varphi; t_0) \equiv \frac{1}{\pi i} \int_{ab} \sqrt{\frac{t-a}{t-b}} \cdot \frac{\Phi_n(\varphi; t, t_0)dt}{t - t_0} =$$

$$= \varphi(t_0) + \frac{1}{\pi i} \int_{ab} H_n(\varphi; t, t_0)dt \qquad\qquad (1.1.3)$$

(In order to obtain the approximate formula, the equality

$$\frac{1}{\pi i} \int_{ab} \sqrt{\frac{t-a}{t-b}} \cdot \frac{dt}{t - t_0} = 1$$

was used, which can be easily proven via Cauchy and Plemelj formulas [77])

The right-hand side of the expression (1.1.3) can be easily offered as

$$p_{\sigma k}^{(1/2;-1/2)} = \frac{1}{\pi i} \int_{ab} \sqrt{\frac{t-a}{t-b}} \cdot l_{\sigma k}(t)dt \,, (\sigma = 1,2,\dots,n; \ = 1,2,\dots,m)$$

with values of function φ at points $t_{\sigma k}$.

As follows from the reasoning above, the integral (1.1.2) is defined when $t_0 \to c$ (c is a or b) and φ satisfies on the line ab (including the endpoints a and b) the Hölder condition with an exponent greater than ½. As will be demonstrated later, this condition is crucial for the numerical solution of singular integral equations that involve integrals of type (1.1.2).

Based on this, define the given integral at points $t_0 = a, b$ in terms of appropriate limit values, we obtain estimates that will be valid for any $t_0 \in ab$, including the endpoints a and b. This latter condition, as will be shown below, is effectively applied to the numerical solution of corresponding singular integral equations (see Chapter II).

We say that $\varphi \in H_\alpha^{(r)}(L)$ if the function φ is continuous on the given smooth contour L, including its derivatives up to r-th order, and at this, r-th derivative $\varphi^{(r)} \in H$ with exponent α $(0 < \alpha \leq 1)$. When $r = 0$, instead of $H_\alpha^{(0)}(L)$ ($½ < \alpha \leq 1$) we will use $H_\alpha(L)$.

From the very beginning we note that due to known justifications ([77] , § 18-22), if $\varphi \in H_\alpha(L)$ $(1/2 < \alpha \leq 1)$ then

$$S_n^{(1/2;-1/2)}(\varphi, t_0) \in H.$$

For any $\varphi \in H_\alpha^{(r)}(L)$ $(1/2 < \alpha \leq 1, \quad r \leq m)$ we can show validity of estimation

$$\left| S^{(1/2;-1/2)}(\varphi, t_0) - S_n^{(1/2;-1/2)}(\varphi, t_0) \right| \leq \frac{C_m \ln n}{n^{r+\alpha-1/2}}, \quad (n > 1)$$

where constant C_m depends only on contour $L \equiv ab$, points $\{x_k\}_{k=1}^m$ and Hölder exponent of function $\varphi^{(r)}$.

Indeed, For the sake of simplicity, we will consider the case $r = 0$ in more detail.

Assume $t_0 \in \tau_\nu \tau_{\nu+1}, t \in \tau_\sigma \tau_{\sigma+1}$ $(\sigma \neq \nu, \sigma \neq \nu \pm 1)$, consider

$$\varphi(t) - \Phi_n(\varphi; t, t_0) =$$

$$= \varphi(t) - \left\{ \varphi(t_0) + \sum_{k=1}^{m} l_{\sigma k}(t) \frac{t - t_0}{t_{\sigma k} - t_0} \varphi(t_{\sigma k}) \right.$$

$$\left. + \sum_{k=1}^{m} l_{\sigma k}(t) \frac{t - t_0}{t_0 - t_{\sigma k}} L_v(\varphi; t_0) \right\}.$$

Take into account a known property of Lagrange interpolation formula

$$\sum_{k=1}^{m} l_{\sigma k}(t) \frac{t - t_0}{t_0 - t_{\sigma k}} = -1 + \frac{\omega_\sigma(t)}{\omega_\sigma(t_0)}. \tag{1.1.5}$$

We can write for any t and $t_0 \neq t_{\sigma k}$

$$\varphi(t) - \Phi_n(\varphi; t, t_0) = \frac{\omega_\sigma(t)}{\omega_\sigma(t_0)} [\varphi(t) - \varphi(t_0)] + \left[1 - \frac{\omega_\sigma(t)}{\omega_\sigma(t_0)} \right] \times$$

$$\times L_v(\varphi - \varphi(t_0); t_0) + \sum_{k=1}^{m} l_{\sigma k}(t) \frac{t - t_0}{t_{\sigma k} - t_0} [\varphi(t) - \varphi(t_{\sigma k})] \tag{1.1.6}$$

$$(t_0 \in \tau_v \tau_{v+1}, t \in \tau_\sigma \tau_{\sigma+1} \; \sigma \neq v, \sigma \neq v \pm 1),$$

$$L_v(\varphi - \varphi(t_0); t_0) = \sum_{k_0=1}^{m} l_{v k_0}(t_0) [\varphi(t_{v k_0}) - \varphi(t_0)] =$$

$$= \sum_{k_0=1}^{m} l_{v k_0}(t_0) \varphi(t_{v k_0}) - \varphi(t_0).$$

If $\sigma = v$ then we can easily see that

$$\varphi(t) - \Phi_n(\varphi; t, t_0) = \varphi(t) - \varphi(t_0) -$$

$$- \sum_{k=1}^{m} l_{\sigma k}(t)(t - t_0) \sum_{\substack{k_0=1 \\ k_0 \neq k}} \prod_{\substack{j=1 \\ j \neq k, k_0}} (t - t_{vj}) \frac{\varphi(t_{v k_0}) - \varphi(t_{vk})}{\omega'_v(t_{v k_0})} \tag{1.1.7}$$

$$(t, t_0 \in \tau_v \tau_{v+1}).$$

Presentations analogous to (1.1.7) for $\varphi(t) - \Phi_n(\varphi; t, t_0)$, when $\sigma = v \pm 1$, will take place.

Taking into account (1.1.6) and (1.1.7) we come to

$$\frac{1}{\pi i}\int_{ab}\sqrt{\frac{t-a}{t-b}}\,\frac{\varphi(t)-\Phi_n(\varphi;t,t_0)}{t-t_0}\,dt =$$

$$= \sum_{\substack{\sigma=1 \\ \sigma\neq v\pm1,v}}^{n}\frac{1}{\pi i}\int_{\tau_\sigma\tau_{\sigma+1}}\sqrt{\frac{t-a}{t-b}}\Bigg(\frac{\omega_\sigma(t)}{\omega_\sigma(t_0)}\frac{\varphi(t)-\varphi(t_0)}{t-t_0}+L_v\big(\varphi(t)-\varphi(t_0)\big);t_0\Bigg)$$

$$\times$$

$$\times\left[1-\frac{\omega_\sigma(t)}{\omega_\sigma(t_0)}\right]\frac{1}{t-t_0}+$$

$$+\sum_{k=1}^{m}l_{\sigma k}(t)\frac{\varphi(t)-\varphi(t_{\sigma k})}{t_{\sigma k}-t_0}\}dt +$$

$$+\frac{1}{\pi i}\int_{\tau_v\tau_{v+1}}\sqrt{\frac{t-a}{t-b}}\Bigg\{\frac{\varphi(t)-\varphi(t_0)}{t-t_0}$$

$$-\sum_{k=1}^{m}l_{\sigma k}(t)\sum_{\substack{k_0=1 \\ k_0\neq k}}^{m}\prod_{\substack{j=1 \\ j\neq k,k_0}}^{m}(t_0-t_{vj})\frac{\varphi(t_{vk_0})-\varphi(t_{vk})}{\omega_v'(t_{vk_0})}\Bigg\}dt +$$

$$+\int_{\tau_{v-1}\tau_v}+\int_{\tau_{v+1}\tau_{v+2}} \tag{1.1.8}$$

where the last two integrals have the same form as that in (1.1.8) along the arc $\tau_v\tau_{v+1}$.

Let's pass to estimation of the expression (1.1.8). Firstly, we want to show that for $t_0\in\tau_v\tau_{v+1}$, $t\in\tau_\sigma\tau_{\sigma+1}$ $\sigma\neq v,\sigma\neq v\pm1$ the following

$$\max_{t\in\tau_\sigma\tau_{\sigma+1}}\left|\frac{\omega_\sigma(t)}{\omega_{\sigma+1}(t_0)}\right|\cdot|\varphi(t)-\varphi(t_0)| = O(n^{-\alpha}), \tag{1.1.9}$$

is true, besides the constant in the right hand side may be taken independent on t_0, σ and v. Indeed, for any v, $t\in\tau_\sigma\tau_{\sigma+1}$ $(\sigma\neq v\pm1,v)$ and k $(1\leq k\leq m)$ we have

$$\max_{t,\sigma}|\omega_\sigma(t)| = O(n^{-m}),\quad \min_{t_0\in\tau_v\tau_{v+1}}|t_0-t_{\sigma k}| > qn^{-1},$$

where the constant $q > 0$ is independent of σ and v. In addition

$$\max_{t_0,t} \left| \frac{t - t_0}{t_0 - t_{\sigma k}} \right| = \max_{t_0,t} \left| 1 + \frac{t_{\sigma k} - t}{t_0 - t_{\sigma k}} \right| = O(1)$$

Uniformly with respect to v and σ. From the last expression the following estimation

$$\max_{t \in \tau_\sigma \tau_{\sigma+1}} \left| \frac{\omega_\sigma(t)}{\omega_{\sigma+1}(t_0)} \right| \cdot |\varphi(t) - \varphi(t_0)| = O(n^{-\alpha}),$$

can be esily obtained, which, via $\varphi \in H_\alpha(L)$, gives (1.1.9), which, in its turn, easily yields

$$\max_{t_0 \in \tau_\sigma \tau_{\sigma+1}} \left| \sum_{\substack{\sigma=1 \\ \sigma \neq v \pm 1, v}}^{n} \frac{1}{\pi i} \int_{\tau_\sigma \tau_{\sigma+1}} \sqrt{\frac{t-a}{t-b}} [\varphi(t) - \varphi(t_0)] \frac{dt}{t - t_0} \right|$$

$$= O\left(\frac{\ln n}{n^{\alpha - 1/2}} \right), \quad (n > 1,\ 1/2 < \alpha \leq 1).$$

Further, using again condition $\varphi \in H_\alpha(L)$ (and smoothness of L) we get:

$$\max_{t_0 \in \tau_\sigma \tau_{\sigma+1}} \left| \sum_{\substack{\sigma=1 \\ \sigma \neq v \pm 1, v}}^{n} L_v(\varphi - \varphi(t_0)); t_0) \frac{1}{\pi i} \int_{\tau_\sigma \tau_{\sigma+1}} \sqrt{\frac{t-a}{t-b}} \left[1 - \frac{\omega_\sigma(t)}{\omega_{\sigma+1}(t_0)} \right] \frac{dt}{t - t_0} \right|$$

$$= O\left(\frac{\ln n}{n^{\alpha - 1/2}} \right),$$

$$(n > 1,\ 1/2 < \alpha \leq 1).$$

Besides, as elementary estimations show,

$$\left| \sum_{\substack{\sigma=1 \\ \sigma \neq v \pm 1, v}}^{n} \sum_{k=1}^{m} \int_{\tau_\sigma \tau_{\sigma+1}} \sqrt{\frac{t-a}{t-b}} \, l_{\sigma k}(t) \frac{\varphi(t) - \varphi(t_{\sigma k})}{t_{\sigma k} - t_0} \, dt \right|$$

$$= O\left(\frac{1}{n^{1/2 + \alpha}} \right) \sum_{\substack{\sigma=1 \\ \sigma \neq v \pm 1, v}}^{n} \frac{1}{|t_{\sigma k} - t_0|} = O\left(\frac{\ln n}{n^{\alpha - 1/2}} \right),$$

$$(1/2 < \alpha \leq 1)$$

uniformly with respect to t_0 and v. Further, again via $\varphi \in H_\alpha(L)$ and smoothness of $L = ab$

$$\left| \int_{\tau_v \tau_{v+1}} \sqrt{\frac{t-a}{t-b}} \, \frac{\varphi(t) - \varphi(t_0)}{t-t_0} \, dt \right| \leq 0 \left(\frac{1}{n^{\alpha-1/2}} \right).$$

The following estimation is also valid

$$\max_{t_0 \in \tau_v \tau_{v+1}} \left| \sum_{k=1}^{m} p_{vk} \sum_{\substack{k_0=1 \\ k_0 \neq k}}^{m} \prod_{\substack{j=1 \\ j \neq k, k_0}}^{m} (t_0 - t_{vj}) \frac{\varphi(t_{vk_0}) - \varphi(t_{vk})}{\omega_v(t_{vk})} \right| = 0 \left(\frac{1}{n^{\alpha+1/2}} \right).$$

Similarly can be estimated the integrals in expression (1.1.8) corresponding to arcs $\tau_{v-1}\tau_v, \ \tau_{v+1}\tau_{v+2}$. By combining the obtained estimates, we get convinced in the validity of (1.1.4) when $r = 0$.

When $r \geq 1$ the corresponding estimates can be obtained by expanding the function $\varphi(t)$ into a Taylor series in the vicinity of the point t_0 when $\sigma = v \neq 1, v$ with the remainder in integral form. For $\sigma \neq v \pm 1, v$ by the following formula (with appropriate values of σ)

$$\varphi(t) = \varphi(t_0) + (t - t_0) \sum_{j=0}^{r-1} \frac{(t - t_0)^j}{j! (t_0 - t)^{j+1}} \int_{\tau_\sigma t_0} (t_0 - u)^j \varphi^{(j+1)}(u) du \ +$$

$$+ \frac{t - t_0}{(r-1)!} \frac{(t - t_\sigma)^r}{(t_0 - t_\sigma)^{r+1}} \int_{\tau_\sigma t_0} (t_0 - u)^{r-1} \varphi^{(r)}(u) du \ -$$

$$- \frac{t - t_0}{(r-1)!} \left\{ \int_{\tau_\sigma t} \frac{(t - \tau)^r + (t - \tau)^{r-1}(t_0 - \tau)}{(t_0 - \tau)^2} \varphi^{(r)}(\tau) d\tau \ - \right.$$

$$\left. - (r+1) \int_{\tau_\sigma t} \frac{(t - \tau)^2 d\tau}{(t_0 - \tau)^{r+2}} \int_{\tau t_0} (t_0 - u)^{r-1} \varphi^{(r)}(u) du \right\}.$$

We can easily be convinced in validity of the last expression using integration by parts. Additionally, we must use the equality $\rho(t) - \Phi_n(\rho; t; t_0) = 0$ which holds for any t, t_0 and $\rho(t)$ is an arbitrary polynomial with degree $\leq m$. The validity of this equality can be also verified directly.

It should be noted that the expression $S_n^{(1/2;-1/2)}(\varphi; t_0)$ involves the value of the function φ at a variable point t_0, so taking this expression as approximate value of the singular integral might not be practical. Therefore, from this perspective, the following approximate formula is entirely acceptable:

$$\frac{1}{\pi i}\int_{ab}\sqrt{\frac{t-a}{t-b}}\,\frac{\varphi(t)dt}{t-t_0} \approx S_n^{*(1/2;-1/2)}(\varphi; t_0),$$

$$t_0 \in \tau_\nu\tau_{\nu+1}, (\nu = 1,2 \dots, n)$$

where

$$S_n^{*(1/2;-1/2)}(\varphi; t_0) = L_\nu(\varphi; t_0) + \frac{1}{\pi i}\int_{ab}\lambda_n(\varphi; t; t_0)dt,$$

It is clear that at the division points $t_{\nu k}$, S_n and S_n^* coinside. Further, as it can be easily noticed, for the following expression

$$\left|\frac{1}{\pi i}\int_{ab}\sqrt{\frac{t-a}{t-b}}\,\frac{\varphi(t)dt}{t-t_0} - S_n^{*(1/2;-1/2)}(\varphi; t_0)\right|$$

under proper coditions on φ the estimation of type (1.1.4) is true.

Similarly it can be proven that for any $\varphi \in H_\alpha^{(r)}(L)$, $(1/2 < \alpha \leq 1, \ r \leq m)$ the estimations

$$\left|S^{(-1/2;-1/2)}(\varphi; t_0) - S_n^{*(-1/2;-1/2)}(\varphi; t_0)\right| = 0\left(\frac{\ln n}{n^{r+\alpha-1/2}}\right)$$

and

$$\left|S^{(1/2;1/2)}(\varphi; t_0) - S_n^{*(1/2;1/2)}(\varphi; t_0)\right| = 0\left(\frac{\ln n}{n^{r+\alpha-1/2}}\right).$$

are valid.

We will write the expression $S_n^{*(1/2;-1/2)}(\varphi; t_0)$ in expanded view (for clarity it is meant thay $x_1 = 0$ and $x_m = 1$):

$$S_n^{*(1/2;-1/2)}(\varphi; t_0) =$$

$$= \left[1 + \sum_{\substack{\sigma=1 \\ \sigma \neq v}}^{n} \sum_{k=1}^{m} \frac{p_{\sigma k}^{*(1/2;-1/2)}}{t_{vj} - t_{\sigma k}} + + \sum_{\substack{k=1 \\ k \neq j}}^{m} \frac{p_{vk}^{*(1/2;-1/2)}}{t_{vj} - t_{vk}} - p_{vj}^{*(1/2;-1/2)} \sum_{k=1}^{m} d_{vk}(t_{vj}) \right] \times$$

$$\times \varphi(t_{vj}) - \sum_{\substack{\sigma=1 \\ \sigma \neq v}}^{n} \sum_{k=1}^{m} \frac{p_{\sigma k}^{*(1/2;-1/2)}}{t_{vj} - t_{\sigma k}} \varphi(t_{\sigma k}) -$$

$$- \sum_{\substack{k=1 \\ k \neq j}}^{m} \frac{p_{vk}^{*(1/2;-1/2)}}{t_{vj} - t_{vk}} \varphi(t_{vk})$$

$$+ p_{vj}^{*(1/2;-1/2)} \sum_{k=1}^{m} d_{vk}(t_{vj}) \varphi(t_{vk}) \qquad (1.1.10)$$

where

$$d_{vk} = \frac{\prod_{\substack{j_0=1 \\ j_0 \neq k,j}}^{m} (t_{vj} - t_v j_0)}{\prod_{\substack{j_0=1 \\ j_0 \neq k}}^{m} (t_{vk} - t_{v j_0})}$$

$$p_{\sigma k}^{(1/2;-1/2)} = \frac{1}{\pi i} \int_{\tau_\sigma \tau_{\sigma+1}} \sqrt{\frac{t-a}{t-b}} \prod_{\substack{j=1 \\ j \neq k}}^{m} \frac{t - t_{\sigma j}}{t_{\sigma k} - t_{\sigma j}} \, dt$$

$$p_{\sigma k}^{*(1/2;-1/2)} = \begin{cases} p_{\sigma 1}^{(1/2;-1/2)} + p_{\sigma-1 m}^{(1/2;-1/2)} & \sigma = 1,2,3, \dots n; \\ p_{\sigma k}^{(1/2;-1/2)} & k = 2,3, \dots, m-1; \sigma = 1,2, \dots n \end{cases}$$

$$(\sigma = 1,2,\cdots n; \quad \kappa = 1,2,\cdots m), \quad (t_0 \in \tau_v \tau_{v+1}; \quad v = 1,2, \cdots, n)$$

From the above reasoning we have

$$\max_{t_0 \in ab} \left| S_n^{(1/2;-1/2)}(\varphi; t_0) \right| \leq M_0 \max_{t \in ab} |\varphi(t)| + M_1 \max_{\sigma,\kappa} |\varphi(t_{\sigma k})| \, n^{1/2} \ln n$$

$$\max_{t_0 \in ab} \left| S_n^{*(1/2;-1/2)}(\varphi; t_0) \right| \leq M_2 \max_{\sigma,} |\varphi(t_{\sigma k})| \, n^{1/2} \ln n \qquad (n > 1), \qquad (1.1.11)$$

where M_0, M_1, M_2 are some constants. Analogously we can write out expressions for quadrature formulas $S_n^{*(-1/2;-1/2)}(\varphi; t_0)$, $S_n^{*(1/2;1/2)}(\varphi; t_0)$:

$$S_n^{*(-1/2;-1/2)}(\varphi; t_0)$$

$$= \left[\sum_{\substack{\sigma=1 \\ \sigma \neq \nu}}^{n} \sum_{k=1}^{m} \frac{p_{\sigma k}^{*(-1/2;-1/2)}}{t_{\nu j} - t_{\sigma k}} + \right.$$

$$+ \sum_{\substack{k=1 \\ k \neq j}}^{m} \frac{p_{\nu k}^{*(-1/2;-1/2)}}{t_{\nu j} - t_{\nu k}} - p_{\nu j}^{*(-1/2;-1/2)} \left. \sum_{k=1}^{m} d_{\nu k}(t_{\nu j}) \right] \varphi(t_{\nu j}) -$$

$$- \sum_{\substack{\sigma=1 \\ \sigma \neq \nu}}^{n} \sum_{k=1}^{m} \frac{p_{\sigma k}^{*(-1/2;-1/2)}}{t_{\nu j} - t_{\sigma k}} \varphi(t_{\sigma \kappa}) -$$

$$- \sum_{\substack{k=1 \\ k \neq j}}^{m} \frac{p_{\nu k}^{*(-1/2;-1/2)}}{t_{\nu j} - t_{\nu k}} \varphi(t_{\nu k}) + p_{\nu j}^{*(-1/2;-1/2)} \sum_{k=1}^{m} d_{\nu k}(t_{\nu j}) \varphi(t_{\nu k}) \qquad (1.1.12)$$

$$S_n^{*(1/2;1/2)}(\varphi; t_0) =$$

$$= \left[t_{\nu j} + \sum_{\substack{\sigma=1 \\ \sigma \neq \nu}}^{n} \sum_{k=1}^{m} \frac{p_{\sigma k}^{*(1/2;1/2)}}{t_{\nu j} - t_{\sigma k}} + \sum_{\substack{k=1 \\ k \neq j}}^{m} \frac{p_{\nu k}^{*(1/2;1/2)}}{t_{\nu j} - t_{\nu k}} - p_{\nu j}^{*(1/2;1/2)} \sum_{k=1}^{m} d_{\nu k}(t_{\nu j}) \right] \varphi(t_{\nu j})$$

$$-$$

$$- \sum_{\substack{\sigma=1 \\ \sigma \neq \nu}}^{n} \sum_{k=1}^{m} \frac{p_{\sigma k}^{*(1/2;1/2)}}{t_{\nu j} - t_{\sigma k}} \varphi(t_{\sigma \kappa}) -$$

$$- \sum_{\substack{k=1 \\ k \neq j}}^{m} \frac{p_{\nu k}^{*(1/2;1/2)}}{t_{\nu j} - t_{\nu k}} \varphi(t_{\nu k}) + p_{\nu j}^{*(1/2;1/2)} \sum_{k=1}^{m} d_{\nu k}(t_{\nu j}) \varphi(t_{\nu k}), \qquad (1.1.13)$$

where

$$d_{\nu \kappa} = \frac{\prod_{\substack{j_0=1 \\ j_0 \neq k,j}}^{m} (t_{\nu j} - t_{\nu j_0})}{\prod_{\substack{j_0=1 \\ j_0 \neq k}}^{m} (t_{\nu \kappa} - t_{\nu j_0})},$$

$$p_{\sigma k}^{(-1/2;-1/2)} = \frac{1}{\pi i} \int_{\tau_\sigma \tau_{\sigma+1}} \frac{1}{\sqrt{(t-a)(t-b)}} \prod_{\substack{j=1 \\ j \neq k}}^{m} \frac{t - t_{\sigma j}}{t_{\sigma k} - t_{\sigma j}} dt$$

$$p_{\sigma k}^{(1/2;1/2)} = \frac{1}{\pi i} \int_{\tau_\sigma \tau_{\sigma+1}} \sqrt{(t-a)(t-b)} \prod_{\substack{j=1 \\ j \neq k}}^{m} \frac{t - t_{\sigma j}}{t_{\sigma k} - t_{\sigma j}} dt$$

$$p_{\sigma k}^{*(-1/2;-1/2)} = \begin{cases} p_{\sigma 1}^{(-1/2;-1/2)} + p_{\sigma-1m}^{(-1/2;-1/2)} & \sigma = 1,2,3,\dots n; \\ p_{\sigma k}^{(-1/2;-1/2)} & k = 2,3,\dots,m-1; \ \sigma = 1,2,\dots n \end{cases}$$

$$p_{\sigma k}^{*(1/2;1/2)} = \begin{cases} p_{\sigma 1}^{(1/2;1/2)} + p_{\sigma-1m}^{(1/2;1/2)} & \sigma = 1,2,3,\dots n; \\ p_{\sigma k}^{(1/2;1/2)} & k = 2,3,\dots,m-1; \ \sigma = 1,2,\dots n \end{cases}$$

$$(\sigma = 1,2,\cdots n; \quad k = 1,2,\cdots m), \quad (t_0 \in \tau_v \tau_{v+1}; \quad v = 1,2,\cdots,n)$$

It should be noted that the sums S_n^* lead to a unified computational scheme for all values of the parameter $t_0 \in L \equiv ab$ (especially when t_0 is arbitrarily close to nodes of the contour $L = ab$ or near the endpoints a and b). Due to these properties, they can be conveniently employed in practical computations.

Below are the tables for approximate calculation of singular integrals using the quadrature formulas derived earlier in (1.1.10), (1.1.12), and (1.1.13), at specific endpoints $t_0 = \pm 1$ when the integration contour is a real segment $[-1; 1]$, for various numbers of division of this segment. (the inner divisions are always fixed $m = 4$). The density $\varphi(t) = t^2$ in its turn plays a critical role in addressing practical mechanics-related problems. For example, in numerical solutions to crack problems (see Chapter IV), it is necessary to calculate the intensity coefficients at these points. This factor, in turn, determines the behavior and direction of crack propagation [82].

Table 1

n	t_0	$S^{(1/2;-1/2)}$	$S_n^{*(1/2;-1/2)}(t^2;t_0)$	$R_n^{(1/2;-1/2)} = S - S_n^*$
5	-1	-0,5	-0,496551	-0,003449
	1	1,5	1,48621	0,01379
10	-1	-0,5	-0,499256	-0,00074424
	1	1,5	1,49405	0,0059532
30	-1	-0,5	-0,499962	-0,0000380375
	1	1,5	1,49893	0,00106505
50	-1	-0,5	-0,49999	-0,0000101642
	1	1,5	1,49951	0,00048789

Table 2

n	t_0	$S^{(-1/2;-1/2)}(t^2;t_0)$	$S_n^{*(-1/2;-1/2)}(t^2;t_0)$	$R_n^{(-1/2;-1/2)} = S - S_n^*$
5	-1	-1	-0,99233	-0,00766977
	1	1	0,969321	0,0306791
10	-1	-1	-0,997012	-0,00298762
	1	1	0,976099	0,023901
30	-1	-1	-0,999507	-0,000492854
	1	1	0,9862	0,0137999
50	-1	-1	-0,999777	-0,000222581
	1	1	0,989316	0,0106839

Table 3

n	t_0	$S^{(1/2;1/2)}(t^2;t_0)$	$S_n^{*(1/2;1/2)}(t^2;t_0)$	$R_n^{(1/2;1/2)} = S - S_n^*$
5	-1	-1,5	-1,49471	-0,00529483
	1	1,5	1,47882	0,0211793
10	-1	-1,5	-1,4987	-0,00129719
	1	1,5	1,48962	0,0103775
30	-1	-1,5	-1,49993	-0,000073049
	1	1,5	1,49795	0,00204537
50	-1	-1,5	-1,49998	-0,0000198499
	1	1,5	1,49905	0,000952797

As it is seen from the Tables, the accuracy of (1.1.10), (1.1.12), (1.1.13) increases with increase of n and reaches 10^{-4} for $n = 50$.

§1.2 Approximate Evaluation of Singular Integrals with Piecewise Differentiable Densities

As previously mentioned, the sums used above $S_n^{*(\pm 1/2;\ \pm 1/2)}(\varphi;t_0)$ can approximate the integral with high accuracy only when the density being approximated is sufficiently smooth across the entire contour. However, in practical applications, there are cases when the integral equations involve densities for which we can only assert that they have only piecewise continuous derivatives. For example, in structural mechanics and planar elasticity problems, the boundary contour of the domain may often be sufficiently smooth. However, the corresponding boundary condition on the right-hand side of the equation, which characterises external stresses or displacements is frequently only piecewise differentiable. In such cases, the solution of the boundary integral equation, as a rule, also belongs to the same class. Similarly, in planar mechanics problems involving cracks in anisotropic materials, analogous situations arise.

It should be noted that calculation of singular integrals with such densities can be carried out by subtracting and adding piecewise polynomial functions to the given density. These polynomials are chosen so that the resulting integral can be evaluated exactly. Furthermore, the resulting expressions retain the same type of singularities as the original density.

On the other hand, implementing this approach may not always be advisable due to the fact that, in numerical solutions of integral equations involving such singular integrals, alongside with desired function, their (one-sided) derivateves will occure.

In this section, we assume that the integration contour in the considered integrals is piecewise smooth. Additionally, we assume that the discontinuity points of the derivatives of the corresponding density function coincide with the angular points of the contour.

Below, we will construct quadrature formulas which, unlike those developed in §1.1, have the property of achieving high degree of accuracy under the condition that the corresponding density is piecewise differtiable continuous derivatives only on portions of the contour (in this case, only on smooth parts of the piecewise smooth contour).

Thus, we consider the following integral

$$S(\varphi; t_0) \equiv \frac{1}{\pi i} \int_L \frac{\varphi(t)dt}{t - t_0}, \qquad t_0 \in L. \qquad\qquad (1.2.1)$$

For clarity we mean that the given (piecewise smooth) contour L is closed.

Regarding the function $\varphi(t)$ we mean that along arcs $C_q C_{q+1} \subset L$ (under arc $C_q C_{q+1} \subset L$ when L is a closed contour, we mean the least arc with endpoints C_q, C_{q+1}, at this, direction from C_q to C_{q+1} must correspond to positive direction along L), where C_j ($j = 1,2,\ldots,p$) are angular points of L. $\varphi(t)$ is continuous including its r-th derivatives, with $\varphi^{(r)}(t)$ satisfying the Hölder condition.

Besides, at the points C_j $(j = 1,2,\ldots,p)$ generally, $\varphi'(C_j - 0) \neq \varphi'(C_j + 0)$. We will denote functions of such class by $H_\alpha^{(r)}(L; C_1, C_2, \ldots, C_p)$.

Devide each arc $C_i C_{i+1} (i = 1,2,\ldots,p)$ into parts by points $t_{i\,\sigma k}$, just as in §1.1. In the given singular integrals $S(\varphi; t_0)$ we will approximate function φ in the following way: let's mean that $\quad _0 \in \tau_{qv}\tau_{qv+1}$ $(t_0 \in C_q C_{q+1})$. Suppose

$$
\varphi(t) \approx
\begin{cases}
\Phi_{qn}(\varphi; t, t_0), & \ldots, & t \in C_q C_{q+1}, \\
\Phi_{q+1n}(\varphi; t, C_{q+1}), & \ldots, & t \in C_{q+1} C_{q+2}, \\
-\,-\,-\,-\,-\,-\,- & -\,-\,- & -\,-\,-\,-\,-\,- \\
\Phi_{p-1n}(\varphi; t, C_{p-1}), & \ldots, & t \in C_{p-1} C_p, \\
\Phi_{1n}(\varphi; t, C_1), & \ldots, & t \in C_1 C_2, \\
-\,-\,-\,-\,-\,-\,- & -\,-\,- & -\,-\,-\,-\,-\,- \\
\Phi_{q-2n}(\varphi; t, C_{q-2}) & \ldots, & t \in C_{q-2} C_{q-1}, \\
\Phi_{q-1n}(\varphi; t, C_q) & \ldots, & t \in C_{q-1} C_q,
\end{cases}
$$

where

$$
\Phi_{qn}(\varphi; t, t_0) = \varphi(t_0) +
$$

$$
+ \sum_{k=1}^{m} \frac{\omega_{q\sigma}(t)}{(t - t_{q\sigma k})\omega'_{q\sigma}(t_{q\sigma k})} \cdot \frac{t - t_0}{t_0 - t_{q\sigma k}} \left[L_{qv}(\varphi; t_0) - \varphi(t_{q\sigma k}) \right]
$$

$$
L_{qv}(\varphi; t_0) = \sum_{k_0=1}^{m} \frac{\omega_{q\sigma}(t_0)}{(t_0 - t_{qv\,k_0})\omega'_{q\sigma}(t_{qv\,k_0})}, \qquad \omega_{q\sigma}(t_0) = \prod_{k=0}^{m} (t - t_{q\sigma k}),
$$

$$
(r \leq m)
$$

$$
t \in \tau_{q\sigma}\tau_{q\sigma+1}, \qquad t_0 \in \tau_{qv}\tau_{qv+1},
$$

$$
\Phi_{in}(\varphi; t, C_i) = \varphi(C_i) +
$$

$$
+ \sum_{k=1}^{m} \frac{\omega_{i\sigma}(t)}{(t - t_{i\sigma k})\omega'_{i\sigma}(t_{i\sigma k})} \cdot \frac{t - C_i}{C_i - t_{i\sigma k}} \left[L_{i,0}(\varphi; C_i) - \varphi(t_{i\sigma k}) \right].
$$

If $t \in C_{q-1} C_q$, $i = q - 1$, then in the expressiuons $\Phi_{in}(\varphi; t, C_i)$, instead of $L_{i,0}(\varphi; C_i)$ we will consider $L_{i,0}(\varphi; C_i) = L_{q-1,n-1}(\varphi; C_q)$, where

$$
L_{i,0}(\varphi; C_i) = \sum_{k_0=1}^{m} \frac{\omega_{i0}(C_i)}{(C_i - t_{i0\,k_0})\omega'_{i0}(t_{i0\,k_0})} \varphi(t_{i0k_0}),
$$

$$L_{q-1,n-1}(\varphi; C_q) = \sum_{k_0=1}^{m} \frac{\omega_{q-1,n-1}(C_q)}{\left(C_q - t_{q-1,n-1,k_0}\right)\omega'_{q-1,n-1}\left(t_{q-1,n-1,k_0}\right)} \varphi(t_{q-1,n-1,k_0})$$

$$(i = 1,2,\ldots,q-1,q+1,\ldots,p).$$

Further, for $t_0 \neq C_q, C_{q+1}$ we suppose

$$S(\varphi; t_0) \approx \frac{1}{\pi i}\int_{C_q C_{q+1}} \frac{\Phi_{qn}(\varphi; t, t_0)dt}{t - t_0} + \frac{1}{\pi i}\int_{C_{q+1}C_{q+2}} \frac{\Phi_{q+1n}(\varphi; t, C_{q+1})dt}{t - t_0} + \cdots$$

$$+ \frac{1}{\pi i}\int_{C_{p-1}C_p} \frac{\Phi_{p-1n}(\varphi; t, C_{p-1})dt}{t - t_0} + \frac{1}{i}\int_{C_1 C_p} \frac{\Phi_{1n}(\varphi; t, C_i)dt}{t - t_0} + \cdots$$

$$+ \frac{1}{\pi i}\int_{C_{q-2}C_{q-1}} \frac{\Phi_{q-2n}(\varphi; t, C_{q-2})dt}{t - t_0} + \frac{1}{\pi i}\int_{C_{q-1}C_q} \frac{\Phi_{q-1n}(\varphi; t, C_q)dt}{t - t_0}. \quad (1.2.2)$$

Now consider singular integrals. We get

$$\frac{1}{\pi i}\int_{C_i C_{i+1}} \frac{\Phi_{in}(\varphi; t, C_i)dt}{t - t_0} = \sum_{\sigma=1}^{n} \frac{1}{\pi i} \times$$

$$\int_{\tau_{i\sigma}\tau_{i\sigma+1}} \left\{ \frac{\varphi(C_i)}{t - t_0} \right.$$

$$+ \sum_{k=1}^{m} \frac{\omega_{i\sigma}(t)}{(t - t_{i\sigma k})\omega'_{i\sigma}(t_{i\sigma k})} \times$$

$$\times \frac{t - C_i}{(C_i - t_{i\sigma k})(t - t_0)} \left[L_{iv}(\varphi; C_i) - \varphi(t_{i\sigma k}) \right] \Big\} dt =$$

$$= \sum_{\sigma=1}^{n} \left[\frac{\varphi(C_i)}{\pi i} \ln \frac{\tau_{i\sigma+1} - t_0}{\tau_{i\sigma} - t_0} + \right.$$

$$+ \sum_{k=1}^{m} \frac{L_{iv}(\varphi; C_i) - \varphi(t_{i\sigma k})}{C_i - t_{i\sigma k}} \times \frac{1}{\pi i}\int_{\tau_{i\sigma}\tau_{i\sigma+1}} \frac{\omega_{i\sigma}(t)}{(t - t_{i\sigma k})\omega'_{i\sigma}(t_{i\sigma k})} \frac{t - C_i}{t - t_0} dt \right],$$

Then

$$\frac{1}{\pi i}\int_{\tau_{i\sigma}\tau_{i\sigma+1}} \frac{\omega_{i\sigma}(t)}{(t - t_{i\sigma k})\omega'_{i\sigma}(t_{i\sigma k})} \frac{t - C_i}{t - t_0} dt =$$

$$= \frac{1}{\pi i}\int_{\tau_{i\sigma}\tau_{i\sigma+1}} \frac{\omega_{i\sigma}(t)}{(t - t_{i\sigma k})\omega'_{i\sigma}(t_{i\sigma k})} \left(1 + \frac{t_0 - C_i}{t - t_0}\right) dt =$$

$$= p_{i\sigma k} + \frac{t_0 - C_i}{\pi i} \times \int_{\tau_{i\sigma}\tau_{i\sigma+1}} \frac{\omega_{i\sigma}(t)}{(t - t_{i\sigma k})\omega'_{i\sigma}(t_{i\sigma k})(t - t_0)} dt =$$

$$= p_{i\sigma k} + \frac{t_0 - C_i}{\pi i(t_{i\sigma k} - t_0)} \times \int_{\tau_{i\sigma}\tau_{i\sigma+1}} \frac{\omega_{i\sigma}(t)}{\omega'_{i\sigma}(t_{i\sigma k})} \left(\frac{1}{t - t_{i\sigma k}} - \frac{1}{t - t_0}\right) dt =$$

$$= p_{i\sigma k} + \frac{(t_0 - C_i)p_{i\sigma k}}{\pi i(t_{i\sigma k} - t_0)} -$$

$$- \left[\frac{t_0 - C_i}{\pi i \, \omega'_{i\sigma}(t_{i\sigma k})(t_{i\sigma k} - t_0)} \left(\int_{\tau_{i\sigma}\tau_{i\sigma+1}} \frac{\omega_{i\sigma}(t) - \omega_{i\sigma}(t_0)}{t - t_0} dt \right.\right.$$

$$\left.\left. + \omega_{i\sigma}(t_0) \ln\frac{\tau_{i\sigma+1} - t_0}{\tau_{i\sigma} - t_0}\right)\right] =$$

$$= p_{i\sigma k} + \frac{(t_0 - C_i)p_{i\sigma k}}{t_{i\sigma k} - t_0} +$$

$$+ \frac{(t_0 - C_i)\omega_{i\sigma}(t_0)}{\omega'_{i\sigma}(t_{i\sigma k})(t_0 - t_{i\sigma k})} \left(\sum_{j=1}^{m} \frac{p_{i\sigma j}}{t_0 - t_{i\sigma j}} + \frac{1}{\pi i} \ln\frac{\tau_{i\sigma+1} - t_0}{\tau_{i\sigma} - t_0}\right) =$$

$$= p_{i\sigma k} + \frac{t_0 - C_i}{\omega'_{i\sigma}(t_{i\sigma k})(t_0 - t_{i\sigma k})} \left[\omega_{i\sigma}(t_0) \left(\sum_{j=1}^{m} \frac{p_{i\sigma j}}{t_0 - t_{i\sigma j}} + \frac{1}{\pi i} \ln\frac{\tau_{i\sigma+1} - t_0}{\tau_{i\sigma} - t_0}\right) - \right.$$

$$\left. - p_{i\sigma k}\omega'_{i\sigma k}(t_{i\sigma k})\right],$$

where

$$p_{i\sigma k} = \frac{1}{\pi i} \int_{\tau_{i\sigma}\tau_{i\sigma+1}} \frac{\omega_{i\sigma}(t)dt}{(t - t_{i\sigma k})\omega'_{i\sigma}(t_{i\sigma k})}$$

$$(i = 1,2,\cdots,q - 1, q + 1,\cdots p), \qquad (t_0 \in C_q C_{q+1}).$$

Taking this into account we get (as it was marked out, if $i = q - 1$, then $L_{i,0}(\varphi; C_i)$ is changed by $L_{q-1,n-1}(\varphi; C_q)$.

Introduce notations:

$$\frac{1}{\pi i} \int_{C_i C_{i+1}} \frac{\Phi_{in}(\varphi; t, C_i)dt}{t - t_0} = \sum_{\sigma=1}^{n} \left\{ \frac{\varphi(C_i)}{\pi i} \ln\frac{\tau_{i\sigma+1} - t_0}{\tau_{i\sigma} - t_0} + \right.$$

$$+ \sum_{k=1}^{m} \frac{L_{i0}(\varphi; C_i) - \varphi(t_{i\sigma k})}{C_i - t_{i\sigma k}} p_{i\sigma k} + \frac{t_0 - C_i}{\omega'_{i\sigma}(t_{i\sigma k})(t_0 - t_{i\sigma k})} \times$$

$$\times \left[\omega_{i\sigma}(t_0) \left(\sum_{j=1}^{m} \frac{p_{i\sigma j}}{t_0 - t_{i\sigma j}} + \frac{1}{\pi i} \ln\frac{\tau_{i\sigma+1} - t_0}{\tau_{i\sigma} - t_0}\right) - p_{i\sigma k}\omega'_{i\sigma k}(t_{i\sigma k})\right] \right\}$$

$$\sum_{j=1}^{m} \frac{p_{i\sigma j}}{t_0 - t_{i\sigma j}} \equiv \lambda_{i\sigma}(t_0), \qquad \frac{1}{\pi i} \ln \frac{\tau_{i\sigma+1} - t_0}{\tau_{i\sigma} - t_0} \equiv \gamma_{i\sigma}(t_0)$$

$$p_{i\sigma k} + \frac{t_0 - C_i}{\omega'_{i\sigma}(t_{i\sigma k})(t_0 - t_{i\sigma k})} \left[\omega_{i\sigma}(t_0)\big(\lambda_{i\sigma}(t_0) + \gamma_{i\sigma}(t_0)\big) - p_{i\sigma k}\omega'_{i\sigma k}(t_{i\sigma k}) \right]$$

$$\equiv W_{i\sigma k}(t_0; C_i)$$

$$(i = 1,2,\dots,q-1,q+1,\cdots,p), \ \ \big(t_0 \in C_q C_{q+1}\big)..$$

Via preceding calculations and notations, for the corresponding approximating sum (which will be denoted by $S_{n,c_1 c_2 \dots, c_p}(\varphi; t_0)$) we obtain

$$S_{n,c_1 c_2 \dots, c_p}(\varphi; t_0) = \sum_{\sigma=1}^{n} \left\{ \varphi(t_0)\gamma_{q\sigma}(t_0) + \sum_{k=1}^{m} \frac{p_{q\sigma k}}{t_0 - t_{q\sigma k}} \left[L_{qv}(\varphi; t_0) - \varphi(t_{q\sigma k}) \right] + \right.$$

$$+ \sum_{\substack{i=1 \\ i \neq q, q-1}}^{p} \left[\varphi(C_i)\gamma_{i\sigma}(t_0) + \sum_{k=1}^{m} \frac{W_{i\sigma k}(t_0, C_i)}{C_i - t_{i\sigma k}} \left[L_{i,0}(\varphi; C_i) - \varphi(t_{i\sigma k}) \right] \right] +$$

$$\left. + \varphi(C_q)\gamma_{q-1\sigma}(t_0) + \sum_{k=1}^{m} \frac{W_{q-1\sigma k}(t_0, C_q)}{C_q - t_{q-1\sigma k}} \left[L_{q-1,n-1}(\varphi; C_q) - \varphi(t_{q-1\sigma k}) \right] \right\} \qquad (1.23)$$

The obtained expressions are valid also for $t_0 \to C_q$ or $t_0 \to C_{q+1}$.

Thus, we can treat the noted approximating expression for $q = \overline{1,p}, \ v = \overline{1,n}$ as defined for any $t_0 \in L$. Our aim is to show that if $\varphi \in H_\alpha^{(r)}(L, C_1, C_2, \cdots, C_p)$, then for any set of points $x_k \in [0,1]$ $\big(k = \overline{1,m}, \ m \geq r\big)$ the following is valid (at any point $t_0 \in L$)

$$\left| S(\varphi; t_0) - S_{n,c_1 c_2 \dots, c_p}(\varphi; t_0) \right| \leq \frac{C \ln n}{n^{r+\alpha}} \quad (n > 1) \qquad (1.2.4)$$

where C is constant independent of n and t_0.

Indeed, estimation of deflection $S(\varphi; t_0) - S_{n,c_1 c_2 \dots, c_p}(\varphi; t_0)$, in fact, is reduced to estimation of the following expression

$$\frac{1}{\pi i} \int_{C_q C_{q+1}} \frac{\varphi(t) - \Phi_{qn}(\varphi; t, t_0)}{t - t_0} \, dt, \qquad \frac{1}{\pi i} \int_{\substack{C_i C_{i+1} \\ i \neq q, q\pm1}} \frac{\varphi(t) - \Phi_{in}(\varphi; t, C_i)}{t - t_0} \, dt,$$

$$\frac{1}{\pi i}\int_{C_{q-1}C_q}\frac{\varphi(t)-\Phi_{q-1n}\big(\varphi;t,C_q\big)}{t-t_0}\,dt,$$

$$\frac{1}{\pi i}\int_{C_qC_{q+1}}\frac{\varphi(t)-\Phi_{q+1n}\big(\varphi;t,C_{q+1}\big)}{t-t_0}\,dt,$$

and this will be carried out below via the estimations given below

$$\left|\frac{1}{\pi i}\int_{C_qC_{q+1}}\frac{\varphi(t)-\Phi_{qn}(\varphi;t,t_0)}{t-t_0}\,dt\right|\leq$$

$$\leq \max_t\left|\varphi(t)-\Phi_{qn}(\varphi;t,t_0)\right|\sum_{\substack{\sigma=1\\ \sigma\neq\nu,\nu\pm1}}^{n}\left|\frac{1}{\pi i}\int_{\tau_{q\sigma}\tau_{q\sigma+1}}\frac{dt}{t-t_0}\right|+$$

$$+\sum_{\sigma=\nu\pm1}\left|\frac{1}{\pi i}\int_{\tau_{q\sigma}\tau_{q\sigma+1}}\frac{\varphi(t)-\Phi_{qn}(\varphi;t,t_0)dt}{t-t_0}\right|.$$

Similarly to reasoning in § 1.1, counting (1.1.6) and (1.1.7), we get easily

$$\left|\frac{1}{\pi i}\int_{C_qC_{q+1}}\frac{\varphi(t)-\Phi_{qn}(\varphi;t,t_0)}{t-t_0}\,dt\right|\leq\frac{M_1\ln n}{n^{r+\alpha}}\qquad(1.2.5)$$

Now we pass to estimation of the following expression

$$\left|\frac{1}{\pi i}\int_{\substack{C_iC_{i+1}\\ i\neq q,q\pm1}}\frac{\varphi(t)-\Phi_{in}(\varphi;t,C_i)}{t-t_0}\,dt\right|$$

Sinse in this case $t-t_0$ is bounded from below by some constant, independent of n, therefore the last estimation is practically reduced to estimation of $|\varphi(t)-\Phi_{in}(\varphi;t,C_i)|$.

Suppose, at first, that $\sigma\neq 0$, then we can present the difference $\varphi(t)-\Phi_{in}(\varphi;t,C_i)$ in the following form

$$\varphi(t)-\Phi_{in}(\varphi;t,C_i)=$$

$$=\varphi(t)-\frac{\omega_{i\sigma}(t)}{\omega_{i\sigma}(C_i)}\varphi(C_i)-\left[1-\frac{\omega_{i\sigma}(t)}{\omega_{i\sigma}(C_i)}\right]\cdot R_{io}(\varphi;C_i)-$$

$$-\sum_{\substack{k=1\\k\neq k_0}}^{m}\frac{\omega_{i\sigma}(t)(t-C_i)\varphi(t_{i\sigma k})}{(t-t_{i\sigma k})\omega'_{i\sigma}(t_{i\sigma k})(t_{i\sigma k}-C_i)},$$

where notation

$$R_{i0}(\varphi;C_i)\equiv\varphi(C_i)-L_{i0}(\varphi;C_i)\qquad(C_i\in\tau_0\tau_1\,,\sigma\neq0).$$

is used.

The expression $\varphi(t)-\Phi_{in}(\varphi;t,C_i)$ vanishes at any $t\in C_iC_{i+1}$ when $\varphi(t)$ is an arbitrary polynomial of degree $\leq m$.

Taking this fact into account, expand function $\varphi(t)$ for any $t\in\tau_{i\sigma}\tau_{i\sigma+1}$ into Taylor series in the neighbourhood of the point $\tau_{i\sigma}$. We have

$$\varphi(t)-\Phi_{in}(\varphi;t,C_i)=\frac{1}{(r-1)!}\{\int_{\tau_{i\sigma}t}(t-u)^{r-1}[\varphi^{(r)}(u)-\varphi^{(r)}(\tau_{i\sigma})]\,du-$$

$$-\frac{\omega_{i\sigma}(t)}{\omega_{i\sigma}(C_i)}\int_{\tau_{i\sigma}C_i}(C_i-u)^{r-1}[\varphi^{(r)}(u)-\varphi^{(r)}(\tau_{i\sigma})]du$$

$$-\left[1-\frac{\omega_{i\sigma}(t)}{\omega_{i\sigma}(C_i)}\right]R_{i0}(\varphi;C_i)-$$

$$-\sum_{k=1}^{m}\frac{\omega_{i\sigma}(t)(t-C_i)}{(t-t_{i\sigma k})\omega'_{i\sigma}(t_{i\sigma k})(t_{i\sigma k}-C_i)}\int_{\tau_{i\sigma}t_{i\sigma k}}(t_{i\sigma k}-u)^{r-1}[\varphi^{(r)}(u)-$$

$$-\varphi^{(r)}(\tau_{i\sigma})]du\}.\qquad(1.2.6)$$

For $\sigma=0$ let's consider

$$\varphi(t)-\Phi_{in}(\varphi;t,C_i)=\varphi(t)-\varphi(C_i)+\sum_{\substack{k=1\\k\neq k_0}}^{m}\frac{\omega_{i\sigma}(t)(t-C_i)}{(t-t_{i\sigma k})\omega'_{i1\sigma}(t_{i\sigma k})(t_{i\sigma k}-C_i)}\times$$

$$\times\sum_{\substack{k_0=1\\k\neq k_0}}^{m}\frac{\omega_{i0}(C_i)}{(C_i-t_{i0k_0})\omega'_{io}(t_{i\sigma k})}[\varphi(t_{i0k_0})-\varphi(t_{i0k})].$$

Analogously, expanding $\varphi(t)$ at $t=C_i$ we get

$$\varphi(t)-\Phi_{in}(\varphi;t,C_i)=\frac{1}{(r-1)!}\{\int_{C_it}(t-u)^{r-1}[\varphi^{(r)}(u)-\varphi^{(r)}(\tau_{i0})]du-$$

$$-\sum_{\substack{k=1\\k\neq k_0}}^{m}\frac{\omega_{i0}(t)(t-C_i)}{(t-t_{i0k})\omega'_{io}(t_{i0k})(C_i-t_{i0k})}\left[\sum_{\substack{k_0=1\\k\neq k_0}}^{m}\frac{\omega_{i0}(C_i)}{(C_i-t_{i0k_0})\omega'_{io}(t_{i0k_0})}\right]$$

$$\int_{C_i t_{i0k_0}}(t_{i0k_0}-u)^{r-1}\left[\varphi^{(r)}(u)-\varphi^{(r)}(\tau_{i0})\right]du-$$

$$-\int_{C_i t_{i0k}}(t_{i0k}-u)^{r-1}\left[\varphi^{(r)}(u)-\varphi^{(r)}(\tau_{i0})\right]du\}\qquad(1.2.7)$$

From representations (1.2.6) and (1.2.7) we have

$$\left|\frac{1}{\pi i}\int_{\substack{C_i C_{i+1}\\i\neq q,q\pm1}}\frac{\varphi(t)-\Phi_{in}(\varphi;t,C_i)}{t-t_0}dt\right|\leq\frac{M_2\ln n}{n^{r+\alpha}}\quad(n>1,M_2=const.).\quad(1.2.8)$$

Now let us estimate

$$\left|\frac{1}{\pi i}\int_{C_{q+1}C_{q+2}}\frac{\varphi(t)-\Phi_{q+1n}\big(\varphi;t,C_{q+1}\big)}{t-t_0}dt\right|\quad t_0\in C_q C_{q+1}.$$

Suppose from the beginning $v\neq n-1$. Via earlier considerations we get easily

$$\left|\frac{1}{\pi i}\int_{C_{q+1}C_{q+2}}\frac{\varphi(t)-\Phi_{q+1n}\big(\varphi;t,C_{q+1}\big)}{t-t_0}dt\right|\leq\frac{M_3\ln n}{n^{r+\alpha}}\quad(M_3=const.).$$

Now suppopose $v=n-1$, then it is clear that

$$\left|\frac{1}{\pi i}\int_{C_{q+1}C_{q+2}}\frac{\varphi(t)-\Phi_{q+1n}\big(\varphi;t,C_{q+1}\big)}{t-t_0}dt\right|\leq$$

$$\leq\left|\sum_{\sigma=1}^{n}\frac{1}{\pi i}\int_{C_{q+1\sigma}C_{q+2\sigma+1}}\frac{\varphi(t)-\Phi_{q+1n}\big(\varphi;t,C_{q+1}\big)}{t-t_0}dt\right|+$$

$$+\left|\frac{1}{\pi i}\int_{C_{q+10}C_{q+11}}\frac{\varphi(t)-\Phi_{q+1n}\big(\varphi;t,C_{q+1}\big)}{t-t_0}dt\right|.$$

Thus, in fact the estimation is reduced to estimation of the last term, since the first term is estimated like the previous case. Use Taylor expansion of $\varphi(t)$ centered at point C_{q+1}

$$\left|\frac{1}{\pi i}\int_{C_{q+10}C_{q+11}}\frac{\varphi(t)-\Phi_{q+1n}(\varphi;t,C_{q+1})}{t-t_0}dt\right|=$$

$$=\frac{1}{(r-1)!}\left|\left\{\frac{1}{\pi i}\int_{\tau_{q+10}\tau_{q+11}}\frac{dt}{t-t_0}\left[\left\{\int_{C_{q+1}t}(t-u)^{r-1}\left[\varphi^{(r)}(u)-\varphi^{(r)}(C_{q+1})\right]du-\right.\right.\right.\right.$$

$$-\sum_{k=1}^{m}\frac{\omega_{q+10}(t)(t-C_{q+1})}{(t-t_{q+10k})\omega'_{q+10}(t_{q+10k})}\right]\times$$

$$\times\sum_{\substack{k_0=1\\k\neq k_0}}^{m}\frac{\omega_{q+10}(C_{q+1})}{(C_{q+1}-t_{q+10k_0})\omega'_{q+10}(t_{q+10k_0})}\left[\int_{C_{q+1}t_{q+10k_0}}(t_{q+10k_0}\right.$$

$$-u)^{r-1}\left[\varphi^{(r)}(u)-\varphi^{(r)}(C_{q+1})\right]\right]du-$$

$$-\int_{C_{q+1}t_{q+10k}}(t_{q+10k}-u)^{r-1}\left[\varphi^{(r)}(u)-\varphi^{(r)}(C_{q+1})\right]\leq$$

$$\leq\frac{M_4}{n^{r+\alpha}}\quad(n>1;\ M_4=const.)$$

Thus

$$\left|\frac{1}{\pi i}\int_{C_{q+1}C_{q+2}}\frac{\varphi(t)-\Phi_{q+1n}(\varphi;t,C_{q+1})}{t-t_0}dt\right|\leq\frac{M_5\ln n}{n^{r+\alpha}}\quad(M_5=const.)\qquad(1.2.9)$$

$$\left|\frac{1}{\pi i}\int_{C_{q-11}C_q}\frac{\varphi(t)-\Phi_{q-1n}(\varphi;t,C_q)}{t-t_0}dt\right|\leq\frac{M_6\ln n}{n^{r+\alpha}}\quad(M_6=const.)\qquad(1.2.10)$$

Combining the obtained estimations (1.2.5), (1.2.8), (1.2.9), and (1.2.10), we get equality (1.2.4).

§1.3 Approximate Calculations of Singular Integrals with Piecewise Power-Type Density Specificity

As above suppose that L is a piece-wise smooth open contour, with endpoints a and b and angular points $q_1, q_2, \ldots, q_p$. We provide that

$$\varphi(t)=\begin{cases}(t-q_1)^{\alpha_1}(t-q_2)^{\beta_1}\varphi_1(t), & t\in q_1 q_2\\(t-q_2)^{\alpha_2}(t-q_3)^{\beta_2}\varphi_2(t), & t\in q_2 q_3\\\cdots\cdots\cdots\cdots\cdots\cdots\cdots\cdots\cdots\cdots\cdots\cdots\cdots\cdots\cdots\cdots\cdots\cdots\\(t-q_{p-2})^{\alpha_{p-1}}(t-q_{p-1})^{\beta_{p-2}}\varphi_{p-2}(t), & t\in q_{p-2} q_{p-1}\\(t-q_{p-1})^{\alpha_{p-1}}(t-q_p)^{\beta_{p-1}}\varphi_{p-1}(t), & t\in q_{p-1} q_p\end{cases}\qquad(1.3.1)$$

$a = q_1$, $b = q_p$, $\varphi_i(t) \in H_{\gamma_i}(q_i, q_{i+1})$ $(\tfrac{1}{2} < \gamma_i \leq 1)$ $(i = 1,2,\cdots p)$ and α_i, β_i $(i = 1,2,\cdots,p)$ take on values $+1/2$ or $-1/2$.

Consider a singular integral of the following type

$$S_{q_1,q_2,\cdots,q_p}^{\left(\vec{\alpha},\vec{\beta}\right)}(\varphi; t_0) = \begin{cases} S^{(\alpha_1,\beta_1)}(\varphi; t_0), & t_0 \in q_1 q_2 \\ S^{(\alpha_2,\beta_2)}(\varphi; t_0), & t_0 \in q_2 q_3 \\ S^{(\alpha_3,\beta_3)}(\varphi; t_0), & t_0 \in q_3 q_4 \\ \cdots\cdots\cdots\cdots\cdots\cdots \\ S^{(\alpha_{p-2},\beta_{p-2})}(\varphi; t_0), & t_0 \in q_{p-2} q_{p-1} \\ S^{(\alpha_{p-1},\beta_{p-1})}(\varphi; t_0), & t_0 \in q_{p-1} q_p \end{cases} \qquad (1.3.2)$$

$$\vec{\alpha} = \left(\alpha_1, \alpha_2, \cdots, \alpha_{p-1}\right), \vec{\beta} = \left(\beta_1, \beta_2, \cdots, \beta_{p-1}\right),$$

where

$$S^{(\alpha_i,\beta_i)}(\varphi; t_0) = \frac{1}{\pi i} \int_{q_i q_{i+1}} (t - q_i)^{\alpha_i} (t - q_{i+1})^{\beta_i} \frac{\varphi_i(t)dt}{t - t_0} \qquad (1.3.3)$$

$$(\varphi_i(t) \in H_{\mu_i}(q_i, q_{i+1}) \ (1/2 < \mu_i \leq 1) \ (i = 1,2,\cdots,p)),$$

$$\rho_i(t) = (t - q_i)^{\alpha_i}(t - q_{i+1})^{\beta_i}, (i = 1,2,\cdots,p),$$

are weight functions under which we mean fixed branches corresponding to these multivalued functions. Under $S_{q_1,q_2,\cdots,q_p}^{\left(\vec{\alpha},\vec{\beta}\right)}(\varphi; q_i)$ we mean the corresponding limit value, at this function $S_{q_1,q_2,\cdots,q_p}^{\left(\vec{\alpha},\vec{\beta}\right)}(\varphi; t_0)$ has first kind discontinuity points, namely, $S_{q_1,q_2,\cdots,q_p}^{\left(\vec{\alpha},\vec{\beta}\right)}(\varphi; q_i^{+}) \neq S_{q_1,q_2,\cdots,q_p}^{\left(\vec{\alpha},\vec{\beta}\right)}(\varphi; q_i^{-})$. In order $S_{q_1,q_2,\cdots,q_p}^{\left(\vec{\alpha},\vec{\beta}\right)}(\varphi; t_0)$ to be defined as a single valued, under $S_{q_1,q_2,\cdots,q_p}^{\left(\vec{\alpha},\vec{\beta}\right)}(\varphi; q_i)$ we always mean that

$$S_{q_1,q_2,\cdots,q_p}^{\left(\vec{\alpha},\vec{\beta}\right)}(\varphi; q_i) = S_{q_1,q_2,\cdots,q_p}^{\left(\vec{\alpha},\vec{\beta}\right)}(\varphi; q_i^{+}).$$

In the role of approximate value of the integral (1.3.2) we take

$$S_{n,\,q_1,q_2,\cdots,q_p}^{\left(\vec{\alpha},\vec{\beta}\right)}(\varphi; t_0) = \begin{cases} S_n^{(\alpha_1,\beta_1)}(\varphi; t_0), & t_0 \in q_1 q_2 \\ S_n^{(\alpha_2,\beta_2)}(\varphi; t_0), & t_0 \in q_2 q_3 \\ S_n^{(\alpha_3,\beta_3)}(\varphi; t_0), & t_0 \in q_3 q_4 \\ \cdots\cdots\cdots\cdots\cdots \\ S_n^{(\alpha_{p-2},\beta_{p-2})}(\varphi; t_0), & t_0 \in q_{p-2} q_{p-1} \\ S_n^{(\alpha_{p-1},\beta_{p-1})}(\varphi; t_0), & t_0 \in q_{p-1} q_p, \end{cases} \tag{1.3.4}$$

where $S_n^{(\alpha_i,\beta_i)}(\varphi; t_0)$ is defined by formula (1.1.10).

We can show that if $\varphi_i \in H_{\mu_i}^{(r)}\ (q_i, q_{i+1})(i = 1,2,\cdots,p)$ then for any $x_k \in [0,1]\ \left(k = \overline{1,m}, m > z + 1\right)$ the following estimation is true at an arbitrary point $t_0 \in L$

$$\left| S_{q_1,q_2,\cdots,q_n}^{\left(\vec{\alpha},\vec{\beta}\right)}(\varphi; t_0) - S_{n,q_q,q_2,\cdots,q_n}^{\left(\vec{\alpha},\vec{\beta}\right)}(\varphi; t_0) \right| \le \frac{A_1 \ln n}{n^{r+\lambda-1/2}},$$
$$(n > 1)(1/2 < \lambda \le 1), \tag{1.3.5}$$

where A_1 is a constant independent of n and t_0.

Indeed, suppose $t_0 \in q_i q_{i+1}$, then

$$\left| S_{q_1,q_2,\cdots,q_n}^{\left(\vec{\alpha},\vec{\beta}\right)}(\varphi; t_0) - S_{n,q_q,q_2,\cdots,q_n}^{\left(\vec{\alpha},\vec{\beta}\right)}(\varphi; t_0) \right|$$
$$= \left| \frac{1}{\pi i} \int_{q_i q_{i+1}} (t - q_i)^{\alpha_i}(t - q_{i+1})^{\beta_i} \frac{\varphi_i(t)}{t - t_0} - S_n^{(\alpha_i,\beta_i)}(\varphi; t_0) \right|$$

Via the reasoning from § 1.1

$$\left| \frac{1}{\pi i} \int_{q_i q_{i+1}} (t - q_i)^{\alpha_i}(t - q_{i+1})^{\beta_i} \frac{\varphi_i(t)}{t - t_0} - S_n^{(\alpha_i,\beta_i)}(\varphi; t_0) \right|$$
$$\le \frac{C_i \ln n}{n^{r+\lambda_i-1/2}} \quad (r \ge 0; 1/2 < \lambda_i \le 1),$$
$$(i = 1,2,\cdots,p).$$

Using notation $A_1 = \max_{1 \le i \le p}\{C_i\},\ \lambda = \min_{1 \le i \le p}\{\lambda_i\}$, we will easily get equality (1.3.5).

§1.4 On Uniform Estimation of Approximation of Singular Integrals with Open Contours Using Piecewise Interpolation Functions

For foundation of various computational processes connected with approximation of singular integrals on open contours, it is crucial to develop methods for uniform estimation of the weighted singular integrals for the entire contour, including its endpoints.

Below, by changing partially the sums $S_n^{(\pm 1/2;\ \pm 1/2)}(\varphi; t_0)$, $S_{n,\ q_1,q_2,\cdots,q_p}^{(\vec{\alpha},\vec{\beta})}(\varphi; t_0)$, we demonstrate that it is possible to approximate singular integrals not only uniformly but also with a norm of Hölder space. Suppose that the system of nodes $\{x_k\}_{k=1}^m$ is closed (i.e. $x_1 = 0, x_m = 1$) in interval $[0,1]$. As in §1.1, introduce points $t_{\sigma k}$ $(\sigma = \overline{1,n}; m = \overline{1,m})$ on $L \equiv ab$ and go back to expressions $S_n^{(\pm 1/2;\ \pm 1/2)}(\varphi; t_0)$ considered in §1.1, which approximate the integrals $S^{(\pm 1/2;\ \pm 1/2)}(\varphi; t_0)$.

For approximation of these integrals, we construct piece-wise interpolation functions $L_n\left[\left(S_n^{(\pm 1/2;\ \pm 1/2)}\varphi\right); t_0\right]$, let us mean

$$L_n\left[\left(S_n^{(\pm 1/2;\ \pm 1/2)}\varphi\right); t_0\right] = L_{n\,j}\left[\left(S_n^{(\pm 1/2;\ \pm 1/2)}\varphi\right); t_0\right],$$
$$t_0 \in \tau_j\tau_{j+1}, \quad \left(j = \overline{1,n}\right),$$

where

$$L_{nj}\left[\left(S_n^{(\pm 1/2;\ \pm 1/2)}\varphi\right); t_0\right] = \sum_{k=1}^m \frac{\varpi_j(t_0)}{(t_0 - t_{jk})\varpi_j(t_{jk})} \cdot S_n^{(\pm 1/2;\pm 1/2)}(\varphi; t_{jk})t_0 \in L_j L_{jn}$$

At this, under $S_n^{(\pm 1/2;\ \pm 1/2)}(\varphi; \tau_j)$ $\left(j = \overline{1,n}\right)$ we mean one of one-sided limits of expression $S_n^{(\pm 1/2;\ \pm 1/2)}(\varphi; t_0)$ at points τ_j.

For definiteness we put

$$S_n^{(\pm 1/2;\ \pm 1/2)}(\varphi; \tau_j) = S_n^{(\pm 1/2;\ \pm 1/2)}(\varphi; \tau_j^+) \quad \left(j = \overline{1,n}\right)$$

It can be easily shown that for any function φ, given on $L \equiv ab$, operator-function $L_n\left[\left(S_n^{(\pm 1/2;\ \pm 1/2)}\varphi\right), t_0\right]$ is continuous on L and, besides, satisfies the Hölder condition (with exponent $\alpha = 1$) on L.

We suppose that

$$S^{(\pm 1/2;\ \pm 1/2)}(\varphi; t_0) \approx L_n\left[\left(S_n^{(\pm 1/2;\ \pm 1/2)}\varphi\right); t_0\right] \quad t_0 \in ab \qquad (1.4.1)$$

It can be shown that in the sums $L_n\left[\left(S_n^{(\pm 1/2;\ \pm 1/2)}\varphi\right); t_0\right]$, dependence of functional coefficients on t_0 takes a considerably simpler form compared to the sums $S_n^{(\pm 1/2;\ \pm 1/2)}(\varphi; t_0)$ discussed in §1.1. This aspect is particularly significant in cases where it is necessary to tabulate the integrals under consideration for a wide range of values of the parameter t_0.

Next, in the estimates provided in §1.1, the right-hand side contains an expression that, while it can be explicitly defined, exhibits sufficiently complex dependence on the singularity parameter t_0.

In the given cases, the corresponding estimates are reduced to expressions of similar type (which is considerably simpler) at the nodal points t_{vk}, along with sufficiently simple estimations for the residual terms of the piecewise interpolating functions.

We introduce a norm on a class of functions $\varphi \in H_\beta(L)$ $(0 < \beta < 1)$:

$$\|\varphi\|_{H_\beta} = \max_{t \in L}|\varphi(t)| + \sup_{t_1, t_2 \in L}\frac{|\varphi(t_1) - \varphi(t_2)|}{|t_1 - t_2|^\beta}.$$

In this case, the considered set becomes a linear normed space, denoted as H_β. As is well known, the space H_β is Banach space.

It is known (see [77], §22) that if $\varphi \in H_\alpha(L)$ $\left(\frac{1}{2} < \alpha \leq 1\right)$ on an open smooth contour $L \equiv ab$, then the singular integral $S^{(\pm 1/2;\ \pm 1/2)}(\varphi; t_0)$, since being a function of t_0, belongs to the Hölder class H with exponent $\alpha - 1/2$. At this $S^{(\pm 1/2;\ \pm 1/2)}(\varphi; t_0)$ assumes a linear operator from $H_\alpha(L)$ into $H_{\alpha - 1/2}(L)$.

From this and the same properties of $L_n\left[\left(S_n^{(\pm 1/2;\ \pm 1/2)}\varphi\right);t_0\right]$ it follows that
$R_n^{(\pm 1/2;\ \pm 1/2)}(S\varphi;t_0)\equiv S^{(\pm 1/2;\ \pm 1/2)}(\varphi;t_0)-L_n\left[S_n^{(\pm 1/2;\ \pm 1/2)}\varphi;t_0\right]\in$
$H_{\alpha-1/2}$, moreover, the noted difference belongs to any class $H_\beta(L),(\beta<\alpha-1/2)$. Consequently, for any $H_\alpha^{(r)}(L)$ $(r\geq 0,\ 1/2<\alpha\leq 1)$ we have
$R_n^{(\pm 1/2;\ \pm 1/2)}(S\varphi;t_0)\in H_\beta(L)$, $(\beta<\alpha-1/2;\ \ 1/2<\alpha\leq 1)$.

Admit that $L\equiv ab$ is an open smooth contour and $\{x_k\}_{k=1}^m$ is any closed nodal system in interval $[0,1]$, then for any $\varphi\in H_\alpha^{(r)}(L)$ $(r\geq 0,\ 1/2<\alpha\leq 1)$ the following estimates are valid:

$$\left\|R_n^{(1/2;\ -1/2)}(S\varphi;t_0)\right\|_{H_\beta}\leq\frac{M_r\ln n}{n^{r+\alpha-\beta-1/2}},\tag{1.4.2}$$

$$\left\|R_n^{(-1/2;\ -1/2)}(S\varphi;t_0)\right\|_{H_\beta}\leq\frac{C_r\ln n}{n^{r+\alpha-\beta-1/2}},\tag{1.4.3}$$

$$\left\|R_n^{(1/2;1/2)}(S\varphi;t_0)\right\|_{H_\beta}\leq\frac{A_r\ln n}{n^{r+\alpha-\beta-1/2}}\tag{1.4.4}$$

$$\left(n>1;\ \beta<\alpha-\frac{1}{2};\ m>r+1\right)$$

Where $M_r,\ C_r,A_r$ are constants dependent only on L and Hölder exponent of function $\varphi^{(r)}$.

Indeed, let us prove (1.4.2). Firstly, we will consider the case $r=0$.

We will show that

$$\max_{t_0\in L}\left|R_n^{(1/2;-1/2)}(S\varphi;t_0)\right|+\sup_{t_1,t_2\in L}\frac{\left|R_n^{(1/2;-1/2)}(S\varphi;t_2)-R_n^{(1/2;-1/2)}(S\varphi;t_1)\right|}{|t_2-t_1|^\beta}=$$
$$O\left(\frac{\ln n}{n^{\alpha-\beta-1/2}}\right)\quad(\beta<\alpha-1/2,\ 1/2<\alpha<1)$$

Firstly, we note that

$$\max_{t_0\in L}\left|R_n^{(1/2;-1/2)}(S\varphi;t_0)\right|=$$
$$=\max_{t_0\in L}\left|S^{(1/2;-1/2)}(\varphi;t_0)-L_n\left[\left(S^{(1/2;-1/2)}\varphi\right);t_0\right]\right.$$
$$\left.+L_n\left[\left(S^{(1/2;-1/2)}-S_n^{(1/2;-1/2)}\varphi\right);t_0\right]\right|\leq$$

$$\leq \max_{t_0 \in L}\left|S^{(1/2;-1/2)}(\varphi;t_0) - L_n\left[\left(S^{(1/2;-1/2)}\varphi\right);t_0\right]\right|$$
$$+ \max_{t_0 \in L}\left|L_n\left[\left(S^{(1/2;-1/2)} - S_n^{(1/2;-1/2)}\varphi\right);t_0\right]\right|,$$

Then

$$\max_{t_0 \in L}\left|S^{(1/2;-1/2)}(\varphi;t_0) - L_n\left[\left(S^{(1/2;-1/2)}\varphi\right);t_0\right]\right| =$$

$$= \max_{\substack{t_0 \in \tau_\nu \tau_{\nu+1} \\ 1 \leq \nu \leq n}} \left|\sum_{k=1}^{m} \frac{\omega_\nu(t_0)}{(t_0 - t_{\nu k})\omega_\nu'(t_{\nu k})}\left[S^{(1/2;-1/2)}(\varphi;t_0) - S^{(1/2;-1/2)}(\varphi;t_{\nu k})\right]\right|.$$

It is evident that

$$\max_{\substack{t_0 \in \tau_\nu \tau_{\nu+1} \\ 1 \leq \nu \leq n}} \left|\sum_{k=1}^{m} \frac{\omega_\nu(t_0)}{(t_0 - t_{\nu k})\omega_\nu'(t_{\nu k})}\right| = O(1),$$

$$\max_{t_0 \in L}\left|S^{(1/2;-1/2)}(\varphi;t_0) - S^{(1/2;-1/2)}(\varphi;t_{\nu k})\right| =$$

$$= \max_{\substack{t_0 \in \tau_\nu \tau_{\nu+1} \\ 1 \leq \nu \leq n}} \left|S^{(1/2;-1/2)}(\varphi;t_0) - S^{(1/2;-1/2)}(\varphi;t_{\nu k})\right| = O\left(\frac{1}{n^{\alpha-1/2}}\right),$$

$$\max_{t_0 \in L}\left|L_n\left[\left(S^{(1/2;-1/2)} - S_n^{(1/2;-1/2)}\varphi\right);t_0\right]\right| =$$

$$= \max_{\substack{t_0 \in \tau_\nu \tau_{\nu+1} \\ 1 \leq \nu \leq n}} \left|\sum_{k=1}^{m} \frac{\omega_\nu(t_0)}{(t_0 - t_{\nu k})\omega_\nu'(t_{\nu k})}\left[\left(S^{(1/2;-1/2)} - S_n^{(1/2;-1/2)}\right)(\varphi;t_{\nu k})\right]\right|.$$

As it was shown in §1.1,

$$\sup_{t_0 \in L}\left|\left(S^{(1/2;-1/2)}\varphi - S_n^{(1/2;-1/2)}\varphi\right)(t_{\nu k})\right| = O\left(\frac{\ln n}{n^{\alpha-1/2}}\right).$$

Therefore

$$\max_{t_0 \in L}\left|L_n\left[\left(S^{(1/2;-1/2)}\varphi - S_n^{(1/2;-1/2)}\varphi\right);t_0\right]\right| = O\left(\frac{\ln n}{n^{\alpha-1/2}}\right).$$

Thus, when $r = 0$ we have

$$\max_{t_0 \in L}\left|R_n^{(1/2;-1/2)}(S\varphi;t_0)\right| = O\left(\frac{\ln n}{n^{\alpha-1/2}}\right).$$

Let us show that in this case the following is true as well:

$$\sup_{t_1,t_2\in L}\frac{\left|R_n^{(1/2;-1/2)}(S\varphi;t_2)-R_n^{(1/2;-1/2)}(S\varphi;t_1)\right|}{|t_2-t_1|^\beta}=O\left(\frac{1}{n^{\alpha-\beta-1/2}}\right)$$

$$(n>1,\beta<\alpha-1/2,\,1/2<\alpha<1).$$

Consider the cases $|t_1-t_2|>n^{-1}$ and $|t_1-t_2|\le n^{-1}$ separately. For $|t_1-t_2|>n^{-1}$ the validity of this inequality is evident. Now suppose $|t_1-t_2|\le n^{-1}$ and let $t_1\in\tau_\nu\tau_{\nu+1},\,t_2\in\tau_\mu\tau_{\mu+1}$ $(t_1\ne t_2)$, we will have

$$\frac{\left|R_n^{(1/2;-1/2)}(S\varphi;t_2)-R_n^{(1/2;-1/2)}(S\varphi;t_1)\right|}{|t_2-t_1|^\beta}=$$

$$=[\{S^{(1/2;-1/2)}(\varphi;t_1)-L_n[(S^{(1/2;-1/2)}\varphi);t_1]+$$

$$+L_n[(S^{(1/2;-1/2)}-S_n^{(1/2;-1/2)}\varphi);t_1]-S^{(1/2;-1/2)}(\varphi;t_2)+$$

$$+L_n[(S^{(1/2;-1/2)}\varphi);t_2]-$$

$$-L_n[(S^{(1/2;-1/2)}-S_n^{(1/2;-1/2)}\varphi);t_2]\}]\cdot\frac{1}{|t_1-t_2|^\beta}\le$$

$$\le\frac{\left|S^{(1/2;-1/2)}(\varphi;t_1)-S^{(1/2;-1/2)}(\varphi;t_2)\right|}{|t_1-t_2|^\beta}+$$

$$+\frac{\left|L_n[(S^{(1/2;-1/2)}\varphi);t_1]-L_n[(S^{(1/2;-1/2)}\varphi);t_2]\right|}{|t_1-t_2|^\beta}+$$

$$+\frac{\left|L_n[(S^{(1/2;-1/2)}\varphi-S_n^{(1/2;-1/2)}\varphi);t_1]-L_n[(S^{(1/2;-1/2)}\varphi-S_n^{(1/2;-1/2)}\varphi);t_2]\right|}{|t_1-t_2|^\beta}$$

At this

$$\frac{\left|S^{(1/2;-1/2)}(\varphi;t_1)-S^{(1/2;-1/2)}(\varphi;t_2)\right|}{|t_1-t_2|^\beta}=O\left(\frac{1}{n^{\alpha-\beta-1/2}}\right)\quad(|t_1-t_2|\le n^{-1}),$$

$$\left|L_n[(S^{(1/2;-1/2)}\varphi);t_1]-L_n[(S^{(1/2;-1/2)}\varphi);t_2]\right|=$$

$$=\left|L_{n\nu}[(S^{(1/2;-1/2)}\varphi);t_1]-L_{n\mu}[(S^{(1/2;-1/2)}\varphi);t_2]\right|=$$

$$=O(1)|t_1-t_2|\sup_{\tau\in L}\left|L_n'\left[\left(S^{\left(\frac{1}{2};-\frac{1}{2}\right)}\varphi\right);\tau\right]\right|,$$

We can represent the noted inequality in the following way as well

$$\left|L_{n\nu}[(S^{(1/2;-1/2)}\varphi);t_1]-L_{n\mu}[(S^{(1/2;-1/2)}\varphi);t_2]\right|=$$

$$=O(1)|t_1-t_2|\sup_{\substack{t\in\tau_j\tau_{j+1}\\1\le j\le n}}\left|L_{nj}'[(\varphi-\varphi(t));t]\right|$$

Via the known property of the Lagrange coefficients, we have

$$\frac{d}{d\tau} L_{nj}(1;\tau) = 0 \quad \left(\tau \in \tau_j \tau_{j+1};\ j = \overline{1,n}\right).$$

Counting this we obtain easily this inequality

$$\left| L_{n\nu}\left[\left(S^{(1/2;-1/2)}\varphi\right); t_1\right] - L_{n\mu}\left[\left(S^{(1/2;-1/2)}\varphi\right); t_2\right]\right| = O\left(n^{3/2-\alpha}\right)|t_1 - t_2|,$$

Also

$$\left| L_n\left[\left(S^{(1/2;-1/2)}\varphi - S_n^{\,(1/2;-1/2)}\varphi\right); t_1\right]\right.$$
$$\left. - L_n\left[\left(S^{(1/2;-1/2)}\varphi - S_n^{\,(1/2;-1/2)}\varphi\right); t_2\right]\right| =$$
$$= O(1)|t_1 - t_2|\sup_{\tau \in L}\left| L_n\left[\left(S^{(1/2;-1/2)}\varphi - S_n^{\,(1/2;-1/2)}\varphi\right); \tau\right]\right| = O(1)\,|t_1 - t_2|$$
$$=$$
$$\sup_{\substack{t_0 \in \tau_\nu \tau_{\nu+1} \\ 1 \leq \nu \leq n}} \left| \frac{d}{d\tau}\left(\sum_{k=1}^{m} \frac{\omega_\nu(\tau)}{(\tau - t_{\nu\,k})\omega_\nu'(t_{\nu\,k})}\right)\left[\left(S^{(1/2;-1/2)}\varphi - S_n^{(1/2;-1/2)}\varphi\right); t_{\nu k}\right]\right|$$
$$= O\left(n^{3/2-\alpha}\right)|t_1 - t_2|.$$

From the above reasoning it follows that

$$\sup_{t_1,t_2 \in L} \frac{\left|R_n^{(1/2;-1/2)}(S\varphi; t_2) - R_n^{(1/2;-1/2)}(S\varphi; t_1)\right|}{|t_2 - t_1|^\beta} = O\left(\frac{1}{n^{\alpha-\beta-1/2}}\right).$$

Thus, for $r = 0$ the estimation (1.4.2) is proven. Now let us assume $r \geq 1$, it is easy to see that for any $n \geq 1$ the function $L_n\left[\left(S^{(1/2;-1/2)}\varphi\right); t_0\right]$ has a piecewise smooth derivative on $L \equiv ab$. According to the differentiation formula for singular integral, the same property holds for the difference for $\varphi \in H_\alpha^{(r)}(L)$ $(r \geq 1)$

$$R_n^{(1/2;-1/2)}(S\varphi; t_0) = S^{(1/2;-1/2)}(\varphi; t_0) - L_n\left[\left(S_n^{(1/2;-1/2)}\varphi\right); t_0\right].$$

On the basis of it, we can use the inequality

$$\frac{1}{|t_1 - t_2|^\beta} \cdot \left|R_n^{(1/2;-1/2)}(S\varphi; t_1) - R_n^{(1/2;-1/2)}(S\varphi; t_2)\right| =$$

$$= O(1)|t_1 - t_2|^{1-\beta} \max_{\substack{t_0 \in \tau_\nu \tau_{\nu+1} \\ 1 \leq \nu \leq n}} \left| \frac{d}{dt_1} R_n^{(1/2;-1/2)}(S\varphi; t_1) \right|$$

Now, we fix any ν $(1 \leq \nu \leq n)$ and estimate

$$\left| \frac{d}{dt_1} R_n^{(1/2;-1/2)}(S\varphi; t_1) \right|.$$

As earlier, here also we will restrict our reasoning by case $|t_1 - t_2| \leq n^{-1}$. We have

$$R_n^{(1/2;-1/2)}(S\varphi; t_0) =$$
$$= \left[S^{(1/2;-1/2)}(\varphi; t_0) - L_n[(S^{(1/2;-1/2)}\varphi); t_0] \right] + L_n\left[\left(S^{(1/2;-1/2)}\varphi - S_n^{(1/2;-1/2)}\varphi \right); t_0 \right],$$
$$\left| \frac{d}{dt_1} R_n^{(1/2;-1/2)}(S\varphi; t_1) \right| \leq \left| \frac{d}{dt_1} \left[S^{(1/2;-1/2)}(\varphi; t_1) - L_n[(S^{(1/2;-1/2)}\varphi); t_1]] \right] \right| +$$
$$+ \left| \frac{d}{dt_1} L_n\left[\left(S^{(1/2;-1/2)}\varphi - S_n^{(1/2;-1/2)}\varphi \right); t_1 \right] \right|$$

Estimate each term separately. Sinse $R_n^{(1/2;-1/2)}(S\varphi; t_0) = 0$ always when φ is an arbitrary polynomial of order $\leq r - 1$, therefore

$$S^{(1/2;-1/2)}(\varphi; t_0) - L_n\left[(S^{(1/2;-1/2)}\varphi); t_0 \right] =$$
$$= \int_{\tau_\nu}^{t_0} (t_0 - u)^{r-1}(S^{(1/2;-1/2)})^{(r)}(u)du -$$
$$- \sum_{k=1}^{m} \frac{\omega_\nu(t_0)}{(t_0 - t_{\nu k})\omega_\nu'(t_{\nu k})} \int_{\tau_\nu}^{t_{\nu k}} (t_{\nu k} - u)^{r-1}(S^{(1/2;-1/2)})^{(r)}(u)du,$$

It is evident from it that

$$\max_{\substack{t_0 \in \tau_\nu \tau_{\nu+1} \\ 1 \leq \nu \leq n}} \left| \frac{d}{dt_1} \left[S^{(1/2;-1/2)}(\varphi; t_1) - L_n[(S^{(1/2;-1/2)}\varphi); t_1] \right] \right| = O\left(\frac{1}{n^{r+\alpha-3/2}} \right).$$

Further, via

$$L_n\left[\left(S^{(1/2;-1/2)}\varphi - S_n^{(1/2;-1/2)}\varphi \right); t_0 \right] =$$

$$= \sum_{k=1}^{m} \frac{\omega_\nu(t_0)}{(t_0 - t_{\nu k})\omega_\nu'(t_{\nu k})} \left[\left(S^{(1/2;-1/2)}\varphi - S_n^{(1/2;-1/2)}\varphi \right); t_{\nu k} \right]$$

we come to

$$\max_{\substack{t_0 \in \tau_\nu \tau_{\nu+1} \\ 1 \le \nu \le n}} \left| \frac{d}{dt_0} \frac{\omega_\nu(t_0)}{(t_0 - t_{\nu k})\omega_\nu'(t_{\nu k})} \right| =$$

$$= O(n) \max_{\substack{t_0 \in \tau_\nu \tau_{\nu+1} \\ 1 \le \nu \le n}} \left| \frac{1}{dt_0} L_n \left[\left(S^{(1/2;-1/2)}\varphi - S_n^{(1/2;-1/2)}\varphi \right); t_0 \right] \right| =$$

$$= O\left(\frac{1}{n^{r+\alpha-3/2}} \right).$$

Thus, for $r \ge 0$ we obtain

$$\left\| R_n^{(1/2;\,-1/2)}(S\varphi; t_0) \right\|_{H_\beta} \le \frac{M_r \ln n}{n^{r+\alpha-\beta-1/2}}$$

$$(n > 1, \beta < \alpha - 1/2,\ 1/2 < \alpha \le 1),$$

Inequalities (1.4.3) and (1.4.4) are proven similarly.

In the same way we can show that for $\varphi_i \in H_{\alpha_i}^{(r)}(q_i q_{i+1})(i = 1,2,\cdots,p)$ the following estimation

$$\left\| S_{q_1,q_2,\cdots,q_p}^{\left(\vec{\alpha},\vec{\beta}\right)}(\varphi; t_0) - L_n \left[S_{n,q_1,q_2,\cdots,q_p}^{\left(\vec{\alpha},\vec{\beta}\right)}(\varphi; t_0), t_0 \right] \right\|_{H_\beta} \le \frac{M_r' \ln n}{n^{r+\alpha-\beta-1/2}}$$

$$(n > 1, \beta < \alpha - 1/2, 1/2 < \alpha \le 1),$$

takes place, where $\alpha = \min_{1 \le i \le p}\{\alpha_i\}$ and M_r' is some constant.

§1.5 On Construction of a Modified Computational Scheme with Increased Order of Accuracy for the Discrete Singularities Method in the Case of Open Contours

In the introduction, works [98–102] were noted, addressing the development and foundation of modified schemes for the method of discrete singularities with increased accuracy for Liapunov smooth contours.

In the mention scheme, for the integral (0, 1), as well as in the case of the classical method of discrete singularities (see [16]), the closed smooth contour L is devided into $2n$ equal parts by two systems of points $\{\tau_j\}_{j=1}^{2n}$.

At this under arc $\tau_m\tau_q$ the least arc of the contour L with endpoints τ_m, τ_q, ordered in the positive direction is meant. Besides, if the index in notation of knot τ_m is negative or overpasses n, the division points with indeces $m \pm 2n$ are meant respectively.

In [99], the approximation

$$\frac{1}{\pi i}\int_L \frac{\varphi(t)dt}{t - t_0} \approx \varphi(t_0) + \sum_{\sigma=1}^{n}(p_{v+2\sigma-1} + p_{v+2\sigma+1})\frac{\varphi(\tau_{v+2\sigma+1}) - \varphi(t_0)}{\tau_{v+2\sigma+1} - t_0}$$

$t_0 \in \tau_{v-1}\tau_{v+1}(t_0 \neq \tau_{v-1}, \tau_{v+1})$, is considered under condition that $t_0 \in \tau_{v-1}\tau_{v+1}(t_0 \neq \tau_{v-1}, \tau_{v+1})$, where

$$p_{v+2\mu-1} = \frac{1}{\pi i}\int_{\tau_{v+2\mu-1}\tau_{v+2\mu+1}} l_{v\mu}^{(0)}(t)dt = \frac{1}{\pi i}\int_{\tau_{v+2\mu-1}\tau_{v+2\mu+1}} l_{v\mu}^{(1)}(t)dt,$$

at this, $l_{v\mu}^{(0)} = \frac{t-\tau_{v+2\mu+1}}{\tau_{v+2\mu-1}-\tau_{v+2\mu+1}}$, $l_{v\mu}^{(1)} = \frac{t-\tau_{v+2\mu-1}}{\tau_{v+2\mu+1}-\tau_{v+2\mu-1}}$. Here the index v corresponds to that number, for which the given value of the parameter t_0 is inclosed in arc $\tau_{v-1}\tau_{v+1}$.

As we can see, the described process for singular integral approximation is based on approximation of the divided difference of $\varphi(t, t_0)$, which is more smooth (in the case of sufficient smoothness of function $\varphi(t)$) than approximating function $\varphi(t)$ in the known classical discrete singularities method (A quadrature formula constructed on this basis, as it can be easily seen, is exact for polynomial $\varphi(t)$ of order ≤ 2).

By assigning specific values $\tau_1, \tau_2, \ldots, \tau_{2n}$ to the parameter t_0, we obtain a certain approximation scheme for the singular integral $(S\varphi)(t_0)$.

In this scheme, unlike the classical approach, the calculational and control nodes (see, e.g., [16]) do not differ from each other once and for all but interchange with each other while passing from an arbitrary value of $t_0 \in \{\tau_j\}_{j=1}^{2n}$ to the nearest knot (an index with a different parity). Overall, in [99], for any given $t_0 \in L$, the approximation of $(S\varphi)(t_0)$ utilizes a piecewise smooth quadratic interpolation constructed based on three consecutive nodes starting from a specific fixed node.

In [99], it is shown that if the function $\varphi(t)$ has a bounded second derivative on L, then the approximation using the mentioned quadratic formula achieves convergence order $O(n^{-2}lnn)$.

If in addition $\varphi(t)$ satisfies the Hölder condition on L with exponent α ($0 < \alpha \leq 1$), then, based on reasonings from [99], we come to conclusion that only the logarithmic factor can be cleared away in estimation $(n^{-2}lnn)$.

We can be assured that the estimation $(n^{-2}lnn)$, with only a slight loss in accuracy, will remain valid even if, instead of using piecewise quadratic approximation, we employ a simpler piecewise linear approximation.

A natural question of constructing analogous schemes for the case of open contours arises.

It is natural that, in this case, additional difficulties arise, particularly due to the behavior of singular integrals at the endpoints of the line (see [77]) and the properties of solutions of the corresponding singular integral equations. This necessitates the approximation of certain weighted singular integrals and their appropriate inclusion in the aforementioned equations.

Let's say we are given a singular integral:

$$S^{(p;q)}(\varphi; t_0) = \frac{1}{\pi i} \int_L (t - a)^p (t - b)^q \frac{\varphi(t)dt}{t - t_0}, \qquad (1.5.1)$$

where $L \equiv ab$ is some open smooth contour with ends a and b on complex plane given parametrically as $t = t(s)$ ($s_a \leq s \leq s_b$). $\rho(t) = (t - a)^p (t - b)^q$,

Is a weght function with parameters p and q either of $\pm\frac{1}{2}$.

Under this multivalued function, we assume a specific fixed branch is taken into consideration.

As mentioned above, numerous practical applied problems (see [16], [77-78], [82]) can be reduced to singular integral equations involving integrals of this type. Therefore, the approximation of such integrals holds great importance, and this topic will be addressed in detail below.

For any natural n, let us divide the interval $[s_a, s_b]$ into $2n$ parts using the points:

$$s_\sigma = \frac{s_b - s_a}{2n}(\sigma - 1), \quad (\sigma = 1,2,\ldots,2n+1)$$

and denote $\tau_\sigma = t(s_\sigma), (\sigma = 1,2,\ldots,2n+1)$. For clearity and simplicity of consideration, we assume that $p = -q = \frac{1}{2}$.

$$S^{\left(\frac{1}{2},-\frac{1}{2}\right)}(\varphi; t_0) = \frac{1}{\pi i}\int_{ab} \sqrt{\frac{t-a}{t-b}}\,\frac{\varphi(t)dt}{t-t_0}$$

$$= \varphi(t_0) + \frac{1}{\pi i}\int_{ab} \sqrt{\frac{t-a}{t-b}}\,\frac{\varphi(t)-\varphi(t_0)}{t-t_0}dt =$$

$$= \varphi(t_0) + \sum_{\sigma=1}^{n}\frac{1}{\pi i}\int_{\tau_{2\sigma-1}\tau_{2\sigma+1}} \sqrt{\frac{t-a}{t-b}}\,\frac{\varphi(t)-\varphi(t_0)}{t-t_0}dt \qquad (1.5.2)$$

Let's perform linear interpolation of the function $\chi(t,t_0) = \frac{\varphi(t)-\varphi(t_0)}{t-t_0}$ at the nodes $\tau_{2\sigma-1}, \tau_{2\sigma+1}$.

$$\chi(t,t_0) \approx \frac{t-\tau_{2\sigma-1}}{\tau_{2\sigma+1}-\tau_{2\sigma-1}}\chi(\tau_{2\sigma+1},t_0) + \frac{t-\tau_{2\sigma+1}}{\tau_{2\sigma-1}-\tau_{2\sigma+1}}\chi(\tau_{2\sigma-1},t_0). \qquad (1.5.3)$$

Let's insert formula (1.5.3) into (1.5.2), we will have

$$S^{\left(\frac{1}{2},-\frac{1}{2}\right)}(\varphi; t_0) = \varphi(t_0) + \sum_{\sigma=1}^{n}\frac{1}{\pi i}\int_{\tau_{2\sigma-1}\tau_{2\sigma+1}} \sqrt{\frac{t-a}{t-b}}\,\frac{\varphi(t)-\varphi(t_0)}{t-t_0}dt \approx$$

$$\approx \varphi(t_0) + \sum_{\sigma=1}^{n} \cdot \frac{1}{\pi i} \int_{\tau_{2\sigma-1}\tau_{2\sigma+1}}^{\cdot} \sqrt{\frac{t-a}{t-b}} \times$$

$$\times \left[\frac{t-\tau_{2\sigma-1}}{\tau_{2\sigma+1}-\tau_{2\sigma-1}} \chi(\tau_{2\sigma-1}, t_0) + \frac{t-\tau_{2\sigma+1}}{\tau_{2\sigma-1}-\tau_{2\sigma+1}} \chi(\tau_{2\sigma+1}, t_0) \right] =$$

$$= \varphi(t_0) + q_1^{\left(\frac{1}{2};-\frac{1}{2}\right)} \chi(\tau_1, t_0) + \sum_{\sigma=1}^{n-1} \left(q1_{2\sigma-1}^{\left(\frac{1}{2};-\frac{1}{2}\right)} + q_{2\sigma+1}^{\left(\frac{1}{2};-\frac{1}{2}\right)} \right) \chi(\tau_{2\sigma+1}, t_0) +$$

$$+ q1_{2n-1}^{\left(\frac{1}{2};-\frac{1}{2}\right)} \chi(\tau_{2n-1}, t_0),$$

where

$$q1_{2\sigma-1}^{\left(\frac{1}{2};-\frac{1}{2}\right)} = \frac{1}{\pi i} \int_{\tau_{2\sigma-1}\tau_{2\sigma+1}}^{\cdot} \sqrt{\frac{t-a}{t-b}} \frac{t-\tau_{2\sigma-1}}{\tau_{2\sigma+1}-\tau_{2\sigma-1}} \, dt,$$

$$q_{2\sigma+1}^{\left(\frac{1}{2};-\frac{1}{2}\right)} = \frac{1}{\pi i} \int_{\tau_{2\sigma-1}\tau_{2\sigma+1}}^{\cdot} \sqrt{\frac{t-a}{t-b}} \frac{t-\tau_{2\sigma+1}}{\tau_{2\sigma-1}-\tau_{2\sigma+1}} \, dt$$

$$(\sigma = 1, 2, \ldots, n).$$

If we choose interpolation points as even-indexed points, then, by analogy with the previous reasoning, we obtain that:

$$D_n^{\left(\frac{1}{2};-\frac{1}{2}\right)}(\varphi; t_0) = \varphi(t_0) + q_2^{\left(\frac{1}{2};-\frac{1}{2}\right)} \chi(\tau_2, t_0) +$$

$$+ \sum_{\sigma=1}^{n-1} \left(q1_{2\sigma}^{\left(\frac{1}{2};-\frac{1}{2}\right)} + q_{2\sigma+2}^{\left(\frac{1}{2};-\frac{1}{2}\right)} \right) \chi(\tau_{2\sigma+2}, t_0) + q1_{2n-2}^{\left(\frac{1}{2};-\frac{1}{2}\right)} \chi(\tau_{2n}, t_0)$$

where

$$q1_{2\sigma}^{\left(\frac{1}{2};-\frac{1}{2}\right)} = \frac{1}{\pi i} \int_{\tau_{2\sigma}\tau_{2\sigma+2}}^{\cdot} \sqrt{\frac{t-a}{t-b}} \frac{t-\tau_{2\sigma}}{\tau_{2\sigma+2}-\tau_{2\sigma}} \, dt,$$

$$q_{2\sigma+2}^{\left(\frac{1}{2};-\frac{1}{2}\right)} = \frac{1}{\pi i} \int_{\tau_{2\sigma}\tau_{2\sigma+2}}^{\cdot} \sqrt{\frac{t-a}{t-b}} \frac{t-\tau_{2\sigma+2}}{\tau_{2\sigma}-\tau_{2\sigma+2}} \, dt$$

$$(\sigma = 1, 2, \ldots, n-1).$$

Quadrature formulas can be constructed similarly:

$$D_n^{\left(\frac{1}{2},\frac{1}{2}\right)}(\varphi; t_0) = t_0\varphi(t_0) + q_1^{\left(\frac{1}{2},\frac{1}{2}\right)}\chi(\tau_1, t_0) +$$

$$+ \sum_{\sigma=1}^{n-1}\left(q1_{2\sigma-1}^{\left(\frac{1}{2},\frac{1}{2}\right)} + q_{2\sigma+1}^{\left(\frac{1}{2},\frac{1}{2}\right)}\right)\chi(\tau_{2\sigma+1}, t_0) + q1_{2n-1}^{\left(\frac{1}{2},\frac{1}{2}\right)}\chi(\tau_{2n+1}, t_0)$$

$$D_n^{\left(-\frac{1}{2},-\frac{1}{2}\right)}(\varphi; t_0) = q_1^{\left(-\frac{1}{2},-\frac{1}{2}\right)}\chi(\tau_1, t_0) +$$

$$+ \sum_{\sigma=1}^{n-1}\left(q1_{2\sigma-1}^{\left(-\frac{1}{2},-\frac{1}{2}\right)} + q_{2\sigma+1}^{\left(-\frac{1}{2},-\frac{1}{2}\right)}\right)\chi(\tau_{2\sigma+1}, t_0) + q1_{2n-1}^{\left(-\frac{1}{2},-\frac{1}{2}\right)}\chi(\tau_{2n+1}, t_0),$$

where

$$q_{2\sigma-1}^{\left(\frac{1}{2},\frac{1}{2}\right)} = \frac{1}{\pi i}\int_{\tau_{2\sigma-1}}^{\tau_{2\sigma+1}}\sqrt{(t-a)(t-b)}\frac{t-\tau_{2\sigma-1}}{\tau_{2\sigma+1}-\tau_{2\sigma-1}}\,dt,$$

$$q_{2\sigma+1}^{\left(\frac{1}{2},\frac{1}{2}\right)} = \frac{1}{\pi i}\int_{\tau_{2\sigma-1}}^{\tau_{2\sigma+1}}\sqrt{(t-a)(t-b)}\frac{t-\tau_{2\sigma+1}}{\tau_{2\sigma-1}-\tau_{2\sigma+1}}\,dt,$$

$$q1_{2\sigma-1}^{\left(-\frac{1}{2},-\frac{1}{2}\right)} = \frac{1}{\pi i}\int_{\tau_{2\sigma-1}}^{\tau_{2\sigma+1}}\frac{1}{\sqrt{(t-a)(t-b)}}\frac{t-\tau_{2\sigma-1}}{\tau_{2\sigma+1}-\tau_{2\sigma-1}})\,dt,$$

$$q_{2\sigma+1}^{\left(-\frac{1}{2},-\frac{1}{2}\right)} = \frac{1}{\pi i}\int_{\tau_{2\sigma-1}}^{\tau_{2\sigma+1}}\frac{1}{\sqrt{(t-a)(t-b)}}\frac{t-\tau_{2\sigma+1}}{\tau_{2\sigma-1}-\tau_{2\sigma+1}}\,dt$$

$$(\sigma = 1,2,\ldots,n).$$

In this case, odd-indexed points are chosen in the role of knots. If we take even-indexed points as the knots, then, accordingly, we obtain

$$D_n^{\left(\frac{1}{2},\frac{1}{2}\right)}(\varphi; t_0) = t_0\varphi(t_0) + q_2^{\left(\frac{1}{2},\frac{1}{2}\right)}\chi(\tau_2, t_0) +$$

$$+ \sum_{\sigma=1}^{n-1}\left(q1_{2\sigma}^{\left(\frac{1}{2},\frac{1}{2}\right)} + q_{2\sigma+2}^{\left(\frac{1}{2},\frac{1}{2}\right)}\right)\chi(\tau_{2\sigma+2}, t_0) + q1_{2n-2}^{\left(\frac{1}{2},\frac{1}{2}\right)}\chi(\tau_{2n}, t_0) \qquad (1.5.4)$$

$$D_n^{\left(-\frac{1}{2},-\frac{1}{2}\right)}(\varphi; t_0) = q_2^{\left(-\frac{1}{2},-\frac{1}{2}\right)}\chi(\tau_2, t_0) +$$

$$+\sum_{\sigma=1}^{n-1}\left(q1_{2\sigma}^{\left(-\frac{1}{2};-\frac{1}{2}\right)}+q_{2\sigma+2}^{\left(-\frac{1}{2};-\frac{1}{2}\right)}\right)\chi(\tau_{2\sigma+2},t_0)+q1_{2n-2}^{\left(-\frac{1}{2};-\frac{1}{2}\right)}\chi(\tau_{2n},t_0),$$

where

$$q_{2\sigma}^{\left(\frac{1}{2};\frac{1}{2}\right)}=\frac{1}{\pi i}\int\limits_{\tau_{2\sigma}\tau_{2\sigma+2}}\sqrt{(t-a)(t-b)}\,\frac{t-\tau_{2\sigma}}{\tau_{2\sigma+2}-\tau_{2\sigma}}\,dt,$$

$$q_{2\sigma+2}^{\left(\frac{1}{2};\frac{1}{2}\right)}=\frac{1}{\pi i}\int\limits_{\tau_{2\sigma}\tau_{2\sigma+2}}\sqrt{(t-a)(t-b)}\,\frac{t-\tau_{2\sigma+2}}{\tau_{2\sigma}-\tau_{2\sigma+2}}\,dt,$$

$$q1_{2\sigma}^{\left(-\frac{1}{2};-\frac{1}{2}\right)}=\frac{1}{\pi i}\int\limits_{\tau_{2\sigma}\tau_{2\sigma+2}}\frac{1}{\sqrt{(t-a)(t-b)}}\frac{t-\tau_{2\sigma}}{\tau_{2\sigma+2}-\tau_{2\sigma}})\,dt,$$

$$q_{2\sigma+2}^{\left(-\frac{1}{2};-\frac{1}{2}\right)}=\frac{1}{\pi i}\int\limits_{\tau_{2\sigma}\tau_{2\sigma+2}}\frac{1}{\sqrt{(t-a)(t-b)}}\frac{t-\tau_{2\sigma+2}}{\tau_{2\sigma}-\tau_{2\sigma+2}}\,dt.$$

$$(\sigma=1,2,\dots,n-1).$$

Through sums $D_n^{\left(\pm\frac{1}{2};\pm\frac{1}{2}\right)}(\varphi;t_0)$ over nodes τ_ν, an approximate formula can be constructed for the corresponding singular integrals $S_n^{\left(\pm\frac{1}{2};\pm\frac{1}{2}\right)}(\varphi;t_0)$, which will be valid for any $t_0\in L$ and will maintain the same order of accuracy as at the nodes τ_ν.

Suppose L_n is a piecewise polynomial interpolation function constructed on the arc $\tau_{2\sigma-1}\tau_{2\sigma+1}$ ($\sigma=1,2,\dots,n$) based on the three consecutive nodes $\tau_{2\sigma-1},\tau_{2\sigma},\tau_{2\sigma+1}$ with the help of corresponding values of D_n.

Denote

$$S^{\left(\pm\frac{1}{2};\pm\frac{1}{2}\right)}(\varphi;t_0)\approx\varphi(t_0)+L_n\left[\left(D_n^{\left(\pm\frac{1}{2};\pm\frac{1}{2}\right)}\varphi\right);t_0\right],\quad t_0\in L, \quad (1.5.5)$$

where as mentioned, L_n, is a piecewise polynomial operator, which represents an interpolation polynomial on each arc whose degree is less than or equal to three.

It is evident that the quadrature formula defined as in (1.5.5) satisfies the above-stated properties. Specifically, it is valid for any $t_0\in L_n$, and on the boundary L, it retains the same order of accuracy which it has at the nodes $\{\tau_\nu\}$.

Let us now proceed to the estimation of the residual term in formula (1.5.5). For the function $\varphi(t)$, we require it to possess a bounded second derivative.

For clarity and to simplify the discussion, let us once again consider a concrete residual term as $\bar{R}_n^{(\frac{1}{2},-\frac{1}{2})}(\varphi; \tau_v)$.

Let us select the odd-indexed division points as the interpolation points and the even-indexed division points as the calculation points.

Firstly, let us estimate $\bar{R}_n^{(\frac{1}{2},-\frac{1}{2})}(\varphi; \tau_v)$. On the basis of construction of the considered quadrature formula we get

$$\bar{R}_n^{(\frac{1}{2},-\frac{1}{2})}(\varphi; \tau_{2v}) = \sum_{\sigma=1}^{n} \bar{R}_{n\sigma}^{(\frac{1}{2},-\frac{1}{2})}(\varphi; \tau_{2v})$$

$$\bar{R}_{n\sigma}^{(\frac{1}{2},-\frac{1}{2})}(\varphi; \tau_{2v}) =$$

$$\frac{1}{\pi i} \int_{\tau_{2\sigma-1}\tau_{2\sigma+1}} \sqrt{\frac{t-a}{t-b}}\,\chi(t,\tau_v)dt - \sum_{k=0}^{1} q_{2\sigma-1k}^{(\frac{1}{2},-\frac{1}{2})}\chi\big(\tau_{2(\sigma+k)-1}, \tau_{2v}\big), \qquad (1.5.6)$$

where

$$q_{2\sigma-10}^{(\frac{1}{2},-\frac{1}{2})} = q_{2\sigma-1}^{(\frac{1}{2},-\frac{1}{2})} = \frac{1}{\pi i} \int_{\tau_{2\sigma-1}\tau_{2\sigma+1}} \sqrt{\frac{t-a}{t-b}}\,\frac{t-\tau_{2\sigma-1}}{\tau_{2\sigma+1}-\tau_{2\sigma-1}}\,dt,$$

$$q_{2\sigma-11}^{(\frac{1}{2},-\frac{1}{2})} = q_{2\sigma+1}^{(\frac{1}{2},-\frac{1}{2})} = \frac{1}{\pi i} \int_{\tau_{2\sigma-1}\tau_{2\sigma+1}} \sqrt{\frac{t-a}{t-b}}\,\frac{t-\tau_{2\sigma+1}}{\tau_{2\sigma-1}-\tau_{2\sigma+1}}\,dt.$$

Evidently, each $\bar{R}_{n\sigma}^{(\frac{1}{2},-\frac{1}{2})}(\varphi; \tau_{2v})$ vanishes when the divided difference $\varphi(t,\tau_v)$ represents a linear function on arc $\tau_{2\sigma-1}\tau_{2\sigma+1}$.

Firstly, we assume that the points t and τ_{2v} are not situated on one and the same arc ($t \in \tau_{2\sigma-1}\tau_{2\sigma+1}$). Let us use the following expansion of $\varphi(t)$

$$\varphi(t) = \varphi(\tau_{2v}) + (t - \tau_{2\sigma-1}) \left\{ \frac{1}{(\tau_{2v} - \tau_{2\sigma-1})} \int\limits_{\tau_{2\sigma-1}}^{\tau_{2v}} \varphi'(u)du + \frac{t - \tau_{2v}}{\tau_{2v} - \tau_{2\sigma-1}} \right.$$

$$\left. \times \int\limits_{\tau_{2\sigma-1}}^{\tau_{2v}} (\tau_{2v} - u)\varphi''(u)\,du \right\} + \frac{(t - \tau_{2v})(t - \tau_{2\sigma-1})^2}{(\tau_{2v} - \tau_{2\sigma-1})^3} \times$$

$$\times \int\limits_{\tau_{2\sigma-1}}^{\tau_{2v}} (\tau_{2v} - u)\,\varphi''(u)du - (t - \tau_v) \times$$

$$\times \int\limits_{\tau_{2\sigma-1}}^{t} \frac{(t - \tau)^2 + (t - \tau)(\tau_{2v} - \tau)}{(\tau_{2v} - \tau)^2} \varphi''(\tau)d\tau - 3\,(t - \tau_{2v}) \times$$

$$\times \int\limits_{\tau_{2\sigma-1}}^{t} \frac{(t - \tau)^2 d\tau}{(\tau_{2v} - \tau)^4} \int\limits_{\tau}^{\tau_{2v}} (\tau_{2v} - u)\,\varphi''(u)du\,, \qquad (1.5.7)$$

The validity of this equality can be easily verified by applying the formulas for integration by parts.

If we consider that the remainder term $\bar{R}_{n\sigma}^{(\frac{1}{2};-\frac{1}{2})}(\varphi; \tau_{2v})$ becomes zero whenever $\chi(t, \tau_{2v})$ is a linear function, we obtain the following:

$$\sum_{\sigma=1}^{n} \bar{R}_{n\sigma}^{(\frac{1}{2};-\frac{1}{2})} = \sum_{\sigma=1}^{n} \left(\frac{1}{\pi i} \int\limits_{\tau_{2\sigma-1}}^{\tau_{2\sigma+1}} \sqrt{\frac{t - a}{t - b}} \frac{(t - \tau_{2\sigma-1})^2}{(\tau_{2v} - \tau_{2\sigma-1})^3} \int\limits_{\tau_{2\sigma-1}}^{\tau_{2v}} (\tau_{2v} - u)\varphi''(u)du \right.$$

$$- \int\limits_{\tau_{2\sigma-1}}^{t} \frac{(t - \tau)^2 + (t - \tau)(\tau_{2v} - \tau)}{(\tau_{2v} - \tau)^2} \varphi''(\tau)d\tau$$

$$\left. - 3 \int\limits_{\tau_{2\sigma-1}}^{t} \frac{(t - \tau)^2 d\tau}{(\tau_{2v} - \tau)^4} \int\limits_{\tau}^{\tau_{2v}} (\tau_{2v} - u)\,\varphi''(u)du \right) dt) -$$

$$- \sum_{k=0}^{1} q_{2\sigma-1k} \left[\frac{\left(\tau_{2(\sigma+k)-1} - \tau_{2\sigma-1}\right)^2}{(\tau_{2v} - \tau_{2\sigma-1})^3} \int\limits_{\tau_{2\sigma-1}}^{\tau_{2v}} (\tau_{2v} - u)\varphi''(u)du \right.$$

$$- \int\limits_{\tau_{2\sigma-1}}^{\tau_{2(\sigma+k)-1}} \frac{\left(\tau_{2(\sigma+k)-1} - \tau\right)^2 + \left(\tau_{2(\sigma+k)-1} - \tau\right)(\tau_{2v} - \tau)}{(\tau_{2v} - \tau)^2} \varphi''(\tau)d\tau$$

$$\left. - 3 \int\limits_{\tau_{2\sigma-1}}^{\tau_{2(\sigma+k)-1}} \frac{\left(\tau_{2(\sigma+k)-1} - \tau\right)^2 d\tau}{(\tau_{2v} - \tau)^4} \int\limits_{\tau}^{\tau_{2v}} (\tau_{2v} - u)\,\varphi''(u)du \right]. \qquad (1.5.8)$$

Now, assume that the points t and τ_{2v} are on the same arc.

Let's use Taylor's expansion:

$$\varphi(t) = \varphi(\tau_{2v}) + (t - \tau_{2v})\varphi'(\tau_{2v}) + \frac{1}{2}\int\limits_{\tau_{2vt}}^{\cdot}(t - u)\,\varphi''(u)du,$$

from which

$$\varphi(t,\tau_{2v}) = \varphi'(\tau_{2v}) + \frac{1}{2(t - \tau_{2v})}\int\limits_{\tau_{2vt}}^{\cdot}(t - u)\,\varphi''(u)du, \tag{1.5.9}$$

$$\bar{R}_{n\sigma}^{\left(\frac{1}{2},-\frac{1}{2}\right)}(\varphi;\tau_{2v}) = \frac{1}{\pi i}\int\limits_{\tau_{2v-1}\tau_{2v+1}}[\sqrt{\frac{t - a}{t - b}}\,\frac{1}{2(t - \tau_{2v})}\int\limits_{\tau_{2vt}}^{\cdot}(t - u)\,\varphi''(u)du]dt -$$

$$-\left[\frac{1}{2(\tau_{2v-1} - \tau_{2+1})}\int\limits_{\tau_{2v}\tau_{2v-1}}^{\cdot}(\tau_{2v-1} - u)\,\varphi''(u)du -\right.$$

$$\left.-\frac{1}{2(\tau_{2v+1} - \tau_{2v})}\int\limits_{\tau_{2v}\tau_{2v+1}}^{\cdot}(\tau_{2v+1} - u)\,du\right] \tag{1.5.10}$$

The evaluation of the sums of integrals in formulas (1.5.8) and (1.5.10) is possible if we pass to an angular abscissa and take into account the smoothness of the line L.

Let's note that $h = l/2n$, where l is the length of the line L, and by considering the corresponding properties of the function φ, on the base of direct evaluation, we can show that for the formulas (1.5.8) and (1.5.10), there is an estimate $0\left(h^{\frac{3}{2}-\varepsilon}lnn\right)$, and based on which we can write:

$$\max_{k}\left|\bar{R}_{n}^{\left(\frac{1}{2},-\frac{1}{2}\right)}(\varphi;\tau_{k})\right| = 0\left(h^{\frac{3}{2}-\varepsilon}\right), \tag{1.5.11}$$

At this we can treat $\varepsilon > 0$ as arbitrarily small value.

Now we can estimate $\bar{R}_{n}^{\left(\frac{1}{2},-\frac{1}{2}\right)}(\varphi;t_0)$ for any $t_0 \in L$. For this, firstly we note that

$$\bar{R}_{n}^{\left(\frac{1}{2},-\frac{1}{2}\right)}(\varphi;t_0) = \bar{S}^{\left(\frac{1}{2},-\frac{1}{2}\right)}(\varphi;t_0) - D_{n}^{\left(\frac{1}{2},-\frac{1}{2}\right)}(\varphi;t_0),$$

where

$$\bar{S}^{\left(\frac{1}{2},-\frac{1}{2}\right)}(\varphi;t_0) = \frac{1}{\pi i}\int_L \sqrt{\frac{t-a}{t-b}}\,\frac{\varphi(t)-\varphi(t_0)}{t-t_0}\,dt.$$

Keep account on the obvious identity

$$\bar{R}_n^{\left(\frac{1}{2},-\frac{1}{2}\right)}(\varphi;t_0)$$
$$= \bar{S}^{\left(\frac{1}{2},-\frac{1}{2}\right)}(\varphi;t_0) - L_n\left[\left(\bar{S}^{\left(\frac{1}{2},-\frac{1}{2}\right)}\varphi\right);t_0\right]$$
$$+ L_n\left[\left(\bar{S}^{\left(\frac{1}{2},-\frac{1}{2}\right)}\varphi - D_n^{\left(\frac{1}{2},-\frac{1}{2}\right)}\varphi\right);t_0\right],$$

where L_n is the above mentioned piece-wise interpolation operator (see (1.5.5)).

Using the rules of differentiation of singular integrals and the known Plemelj-Privalov theorem, we get

$$\max_{t_0\in L}\left|\bar{S}^{\left(\frac{1}{2},-\frac{1}{2}\right)}(\varphi;t_0) - L_n\left[\left(^{-\left(\frac{1}{2},-\frac{1}{2}\right)}\varphi\right);t_0\right]\right| = 0\left(h^{\frac{3}{2}-\varepsilon}\right), \tag{1.5.12}$$

where $\varepsilon > 0$ is an arbitrary smsll positive number. For $L_n\left[\left(\bar{S}^{\left(\frac{1}{2},-\frac{1}{2}\right)}\varphi - D_n^{\left(\frac{1}{2},-\frac{1}{2}\right)}\varphi\right);t\right]$ via

$$L_n\left[\left(\bar{S}^{\left(\frac{1}{2},-\frac{1}{2}\right)}\varphi - D_n^{\left(\frac{1}{2},-\frac{1}{2}\right)}\varphi\right);t_0\right]=0(1)\max_{v}\left|\bar{R}_n^{\left(\frac{1}{2},-\frac{1}{2}\right)}(\varphi;\tau_v)\right|$$ and (1.5.11) an estimation of type (1.5.12) is true. Finally, from this we obtain similar estimation for $\max_{t_0\in L}\left|\bar{R}_n^{\left(\frac{1}{2},-\frac{1}{2}\right)}(\varphi;t_0)\right|$. The same estimations will take place for

$$\max_{t_0\in L}\left|\bar{R}_n^{\left(\frac{1}{2},\frac{1}{2}\right)}(\varphi;t_0)\right|,\ \max_{t_0\in L}\left|\bar{R}_n^{\left(-\frac{1}{2},-\frac{1}{2}\right)}(\varphi;t_0)\right|.$$

Thus, based on the assumptions on function φ for integrals $S^{(\pm\frac{1}{2};\pm\frac{1}{2})}(\varphi;t_0)$, the developed approximation schemes provide a guarantee for the rate of convergence of the process, which is arbitrarily close to the approximation $O\left(h^{\frac{3}{2}}\right)$ $(h \to 0)$.

Like in paragraph 1.1.3, it can be proven that when $L \equiv ab$ is a smooth open contour, for any $\varphi \in H_\alpha^{(2)}(L)$, $(\frac{1}{2} < \alpha \le 1)$, the estimations hold

$$\left\|R_n^{(\frac{1}{2};-\frac{1}{2})}(\varphi;t_0)\right\|_{H_\beta} \le \frac{M \ln n}{n^{\frac{3}{2}+\alpha-\beta}}\left(n > 1, \beta < \alpha - \frac{1}{2}\right), \tag{1.5.12}$$

$$\left\|R_n^{(-\frac{1}{2};-\frac{1}{2})}(\varphi;t_0)\right\|_{H_\beta} \le \frac{A \ln n}{n^{\frac{3}{2}+\alpha-\beta}}\left(n > 1, \beta < \alpha - \frac{1}{2}\right), \tag{1.5.13}$$

$$\left\|R_n^{(\frac{1}{2};\frac{1}{2})}(\varphi;t_0)\right\|_{H_\beta} \le \frac{B \ln n}{n^{\frac{3}{2}+\alpha-\beta}}\left(n > 1, \beta < \alpha - \frac{1}{2}\right), \tag{1.5.14}$$

where M, A, B are constants, dependent only on L and Hölder index of $\varphi^{(2)}$.

R e m a r k. When approximating integral $\int_{\tau_{2\sigma-1}\tau_{2\sigma+1}}^{\cdot} \sqrt{\frac{t-a}{t-b}}, \chi(t,t_0)dt$ we used linear interpolation using points $\tau_{2\sigma-1}, \tau_{2\sigma+1}$ on arc $\tau_{2\sigma-1}\tau_{2\sigma+1}$.

In the case where approximation needs to be increased on arc $\tau_{2\sigma-1}\tau_{2\sigma+1}$, an increased higher-order interpolation should be used. Specifically, the points $\tau_{2\sigma-1}, \tau_{2\sigma}, \tau_{2\sigma+1}$ should be taken as nodes, and then, if $\varphi(t)$ has needed smoothness, the convergence order will increase by one.

Below are tables of values of weighted singular integrals of the type (1.5.1), calculated with the help of quadrature formulas we constructed above, for function $\varphi(t) = t^2$, with the weight $\rho(t)$ and different number of divisions n of the integration contour L, when the last represents a sinusoidal curve within the interval $[-\pi, \pi]$.

Table 4 $\rho(t) = \sqrt{(\pi + t)/(\pi - t)}$

n	t_0	$S^{(\frac{1}{2};-\frac{1}{2})}(\varphi;t_0)$	$S_n^{(\frac{1}{2};-\frac{1}{2})}(\varphi;t_0)$	$R_n^{(\frac{1}{2};-\frac{1}{2})} = S^{(\frac{1}{2};-\frac{1}{2})} - D_n^{(\frac{1}{2};-\frac{1}{2})}$

n	t_0			
5	-2.51327-0.47558i	-19.3329+13.3307i	-19.00531+13.24125i	0.3276+0.0894i
	2.51327+0.47558i	4.3144+55.9428i	4.594+55.8533i	-0.2796+0.895i
10	-2.827-0.2938i	-25.6417+19.371i	-25.5499+19.3167i	0.0918+0.0643i
	2.827+0.29389i	15.1995+61.3228i	15.2914+61.2685i	-0.0919+0.543i
20	-2.9845-0.154508i	-28.5052+24.5582i	-28.479+24.5419i	0.0262+0.0163i
	2.9845+0.154508i	22.70104+62.2644i	22.7365+62.248i	-0.03546+0.0164i

Table 5 $\rho(t) = 1/\sqrt{(\pi + t)(\pi - t)}$

n	t_0	$S^{\left(-\frac{1}{2};-\frac{1}{2}\right)}(\varphi; t_0)$	$S_n^{\left(-\frac{1}{2};-\frac{1}{2}\right)}(\varphi; t_0)$	$R_n^{\left(-\frac{1}{2};-\frac{1}{2}\right)} = S^{\left(-\frac{1}{2};-\frac{1}{2}\right)} - D_n^{\left(\frac{1}{2};-\frac{1}{2}\right)}$
5	-2.51327-0.4755i	-2.39027+11.0252i	-2.30122+10.9967i	0.08905+0.0252i
	2.51327+0.4755i	-2.39027+11.0252i	-2.30122+10.9967i	0.08905+0.0252i
10	-2.82743-0.293i	-1.66192+12.8428i	-1.63269+12.8255i	0.02923+0.0173i
	2.82743+0.2938i	-1.66192+12.8428i	-1.63269+12.8255i	0.02923+0.0173i
20	-2.98451-0.1545i	-0.922265+13.818i	-0.913948+13.813i	0.00832+0.0051i
	2.98451+0.1548i	-0.922265+13.818i	-0.913948+13.813i	0.00832+0.0051i

Table 6 $\rho(t) = \sqrt{(\pi + t)(\pi - t)}$

n	t_0	$S^{\left(\frac{1}{2};\frac{1}{2}\right)}(\varphi; t_0)$	$S_n^{\left(\frac{1}{2};\frac{1}{2}\right)}(\varphi; t_0)$	$R_n^{\left(\frac{1}{2};\frac{1}{2}\right)} = S^{\left(\frac{1}{2};\frac{1}{2}\right)} - D_n^{\left(\frac{1}{2};\frac{1}{2}\right)}$
5	-2.51327-0.4755i	-43.1753+13.1157i	-43.6396+12.6023i	-0.5643+0.5134i
	2.51327+0.4755i	-43.1753+13.1157i	-43.6396+12.6023i	-0.5643+0.5134i
10	-2.82743-0.2938i	-67.9758+24.9156i	-68.1006+24.7515i	-0.1248+0.1641i
	2.82743+0.2938i	-67.9758+24.9156i	-68.1006+24.7515i	-0.1248+0.1641i
20	-2.98451-0.1545i	-82.6162+39.6284i	-82.6482+39.5851i	-0.0198+0.0433i
	2.98451+0.1545i	-82.6162+39.6284i	-82.6482+39.5851i	-0.0198+0.0433i

As shown in Tables 4, 5, and 6, the aforementioned quadrature formulas have a fairly high order of accuracy. At the endpoints of the contour (in the special points), for $n = 20$ the approximation error already reaches the order 10^{-2}.

Clearly, if higher accuracy is required, the number of divisions of the contour should be increased.

Chapter II. Numerical Solutions of Singular Integral Equations

This chapter demonstrates the application of the approximate formulas developed in Chapter I for the numerical solution of a certain class of singular integral equations.

This class of equations encompasses various applied problems (including those in structural mechanics), as well as Dirichlet (correspondingly torsion) and basic boundary value problems whose corresponding integral equations are formulated in terms of Cauchy kernel singular integrals.

§2.1 On Numerical Solution of First-Kind Singular Integral Equations for Open Contours

In this section we consider numerical solution of the first-kind singular integral equation

$$\frac{1}{\pi i}\int_L \frac{\varphi_0(t)dt}{t - t_0} + \frac{1}{\pi i}\int_L K(t_0, t)\varphi_0(t)dt = f(t_0), \qquad (t_0 \in L) \qquad (2.1.1)$$

with open contours under special assummptions on function class of the solution sought.

To such equations are reduced such important problems, as, e.g. Dirichlet plane problem with cuts ([77], §107) (correspondingly, problem of construction of conformally mapping function, problems of torsion of a beam and a rod), stationary temperature problem of a plane with arbitrarily situated cracks ([78], ch. VII), also, basic boundary problems of theory of elasticity.

Many authors have studied various schemes for numerical solution of such equations, including more complex cases when the index of the equation is a negative number.

In this regard, it is particularly noteworthy to mention the results of A. Jishkariani [28-30], I. Lifanov [66-68], and B. Musaev [73-76]. A review of the findings by M. Lavrentiev, A. Kalandiya, and I. Efremov is provided in the work

[22]. In works [87], [103], and [67-68], the foundation of the so-called discrete singularities method is presented for equations of the form (2.1.1). In works [120, 118, 119], various methods for approximate solutions of such equations are discussed. In spite of this, the mentioned works discuss the restrictions on kernel of singular integrals and integration line. A segment of real axis is taken in the role of integration line.

In the case of any contour, transformation of the equation to an equation involving line segment is conceptually possible, but often quite impractical, especially for equations that are related to practical applications (see, for example, [82]). Such transformations are often difficult, because in some cases (for instance, in the problems of theory of cracks), we are more frequently forced to use specific initial conditions.

Another, the most important issue is the convergence rate of the numerical schemes.

In the given section, the scheme developed is used for solving first-order singular integral equation, in the case of any open contour, with the assumption that the solution to the given exact equation has a unique solution within the considered class of solutions.

Under conditions on the right hand side and the kernel (see below), with respect to the given equation, we can obtain sufficiently high convergence order.

Construction of the corresponding algorithms is based on the approximation of singular integrals discussed in §1.1 and rely on the results obtained in §1.4. Based on the result presented in this paragraph, where the convergence of quadrature processes is proven assuming open contours, in the metric H_β, it is possible to validate the corresponding calculation process.

In additions, the used quadratic processes allow us to easily develop corresponding real calculation schemes.

Thus, we assume that in the equation (2.1.1), $L \equiv ab$ represents a smooth open contour, which we will again consider to be given in parametric form $t = t(s)$ $(s_a \leq s \leq s_b)$. $K(t_0, t)$ and $f(t)$ are functions given on L, and satisfying the Hölder condition (see [77]).

We want to find a solution to equation (2.1.1), which is bounded at the end-point a and unbounded at b. Such solution, as known (see [77], §86), has the following form:

$$\varphi_0(t) = \sqrt{\frac{t-a}{t-b}}\,\varphi(t),$$

Where the function satisfies Hölder condition on L provided assumptions on K and f. By reversing the characteristic part, the given equation can be derived in the form (see [26] § 42)

$$\varphi(t_0) + \frac{1}{\pi i}\int_{ab}\left[\frac{1}{\pi i}\int_{ab}\sqrt{\frac{t_1-b}{t_1-a}}\,\frac{K(t_1,t)}{t_1-t_0}\,dt_1\right]\varphi_0(t)dt =$$

$$\frac{1}{\pi i}\int_{ab}\sqrt{\frac{t-b}{t-a}}\,\frac{f(t)}{t-t_0}\,dt \qquad (2.1.2)$$

Further we will treat this equation as an operator one. In this connection we will represent it as follows

$$K^0\varphi = I\varphi + k\varphi = f_0(t_0), \qquad (2.1.3)$$

where

$$k\varphi = \frac{1}{\pi i}\int_{ab}\left[\frac{1}{\pi i}\int_{ab}\sqrt{\frac{t_1-b}{t_1-a}}\,\frac{K(t_1,t)}{t_1-t_0}\,dt_1\right]\varphi_0(t)dt$$

$$f_0(t) = \frac{1}{\pi i}\int_{ab}\sqrt{\frac{t-b}{t-a}}\,\frac{f(t)}{t-t_0}\,dt$$

and I is a unitary operator.

As we can see, the free term and the kernel of the equation represent singular integrals

$$(k\varphi)(t_0;t) = \frac{1}{\pi i}\int_{ab}\left[\frac{1}{\pi i}\int_{ab}\sqrt{\frac{t_1-b}{t_1-a}}\,\frac{K(t_1,t)}{t_1-t_0}\,dt_1\right]\varphi_0(t)dt$$

Further, replacing in the integral operator the following term

$$\frac{1}{\pi i}\int_{ab}\sqrt{\frac{t_1-b}{t_1-a}}\,\frac{K(t_1,t)}{t_1-t_0}\,dt_1$$

by quadrature formula, constructed by us in §1.4, we get

$$\frac{1}{\pi i}\int_{ab}\sqrt{\frac{t_1-b}{t_1-a}}\,\frac{K(t_1,t)}{t_1-t_0}\,dt_1\approx L_n\left[\left(S_n^{(1/2;-1/2)}\right);t_0,t\right],\qquad t_0\in ab$$

where

$$L_{nj}\left[\left(S_n^{(1/2;-1/2)}K\right);t_0,t\right]=\sum_{k=1}^{m}\frac{\omega_j(t_0)}{(t_0-t_{jk})\omega_j'(t_{jk})}\left(S_n^{(1/2;-1/2)}K;t_{jk};t\right),$$

$$t_0\in\tau_j\tau_{j+1}$$

similarly

$$\frac{1}{\pi i}\int_{ab}\sqrt{\frac{t-a}{t-b}}\,\frac{f(t)dt}{t-t_0}\approx L_n\left[\left(S^{(1/2;-1/2)}f\right);t_0\right].$$

Let us consider the equation

$$K_n^0\left[L_n[\varphi_n;t_0]\right]=L_n[\varphi_n;t_0]+\frac{1}{\pi i}\int_{ab}L_n\left[(S_nK)^{(1/2;-1/2)};t_0,t\right]L_n[\varphi_n;t]dt=$$

$$=L_n\left[\left(S_n^{(1/2;-1/2)}f\right);t_0\right]\qquad\qquad(2.1.3')$$

in the subspace of functions $L_n[\varphi_n;t_0]$, together with (2.1.2). If in this equation we assign a value from the set

$$T(v;j)=\{t_{vj}\}\quad(v=1,2,\dots n;j=1,2,\dots m)$$

to the parameter t_0, we will get a system of linear equations at points $T(v;j)=\{t_{vj}\}\quad(v=1,2,\dots n;j=1,2,\dots m)$

with unknowns $\varphi_n(t_{vj})$:

$$\left[K_n^0[L_n[\varphi_n; t_0]]\right]_{t_0=t_{vj}} = \left[L_n\left[\left(S_n^{1/2;-1/2}f\right); t_0\right]\right]_{t_0=t_{vj}} \qquad (v = \overline{1,n}; j = \overline{1,m}).$$

More specifically, this system will take the following form

$$\varphi_n(t_{vj}) + \sum_{i=1}^{n}\sum_{e=0}^{m} q_{ie}A_n(t_{ie}, t_{vj})\varphi_n(t_{ie}) = f_0(t_{vj}), \qquad (2.1.3'')$$

where

$$A_n(t_{ie}, t_{vj}) =$$

$$= \left[1 + \sum_{\substack{\sigma=1\\\sigma\neq v}}^{n}\sum_{k=1}^{m}\frac{P_{\sigma k}^{*(1/2;-1/2)}}{t_{\sigma k} - t_{vj}} + \sum_{\substack{k=1\\k\neq j}}^{m}\frac{P_{vk}^{*(1/2;-1/2)}}{t_{vk} - t_{vj}}\right.$$

$$\left. - P_{vj}^{*(1/2;-1/2)}\sum_{\substack{k=1\\k\neq j}}^{m} d_{vk}(t_{vj})\right]K(t_{vj}; t_{ie}) -$$

$$- \sum_{\substack{\sigma=1\\\sigma\neq v}}^{n}\sum_{k=1}^{m}\frac{P_{\sigma k}^{*(1/2;-1/2)}}{t_{\sigma k} - t_{vj}}K(t_{\sigma k}; t_{ie}) - \sum_{\substack{k=1\\k\neq j}}^{m}\frac{P_{vk}^{*(1/2;-1/2)}}{t_{vk} - t_{vj}}K(t_{vk}; t_{ie}) +$$

$$+ P_{vj}^{*(1/2;-1/2)}\sum_{\substack{k=1\\k\neq j}}^{m} d_{vk}(t_{vj})K(t_{vk}; t_{ie}),$$

and $\varphi_n(t_{vj})$ are quantities sought.

$$f_0(t_{vj}) = \left[1 + \sum_{\substack{\sigma=1\\\sigma\neq v}}^{n}\sum_{k=1}^{m}\frac{P_{\sigma k}^{*(1/2;-1/2)}}{t_{\sigma k} - t_{vj}} + \sum_{\substack{k=1\\k\neq j}}^{m}\frac{P_{vk}^{*(1/2;-1/2)}}{t_{vk} - t_{vj}}\right.$$

$$\left. - P_{vj}^{*(1/2;-1/2)}\sum_{\substack{k=1\\k\neq j}}^{m} d_{vk}(t_{vj})\right]f(t_{vj}) -$$

$$\sum_{\substack{\sigma=1\\\sigma\neq v}}^{n}\sum_{k=1}^{m}\frac{P_{\sigma k}^{*(1/2;-1/2)}}{t_{vj} - t_{\sigma k}}f(t_{\sigma k}) - \sum_{\substack{k=1\\k\neq j}}^{m}\frac{P_{vk}^{*(1/2;-1/2)}}{t_{vj} - t_{vk}}f(t_{vk}) +$$

$$+P_{vj}^{*(1/2;-1/2)} \sum_{\substack{k=1 \\ k \neq j}}^{m} d_{vk}(t_{vk}) f(t_{vj})$$

$$(v = 1,2,\ldots,n; \quad j = 1,2,\ldots,m)$$

$$P_{\sigma k}^{(1/2;-1/2)} = \frac{1}{\pi i} \int_{\tau_\sigma \tau_{\sigma+1}} \sqrt{\frac{t-b}{t-a}} \prod_{\substack{j=1 \\ k \neq j}}^{m} \frac{t - t_{\sigma j}}{t_{\sigma k} - t_{\sigma j}} \, dt$$

$$q_{ij}^{(-1/2;1/2)} = \frac{1}{\pi i} \int_{\tau_i \tau_{i+1}} \sqrt{\frac{t-a}{t-b}} \prod_{\substack{j=1 \\ k \neq j}}^{m} \frac{t - t_{ij}}{t_{ie} - t_{ij}} \, dt$$

$$P_{\sigma k}^{*(1/2;-1/2)} = \begin{cases} P_{\sigma k}^{(1/2;-1/2)}, & k = \overline{2, m-1}; \sigma = \overline{1, n}, \\ P_{\sigma 1}^{*(1/2;-1/2)} + P_{\sigma-1,m}^{*(1/2;-1/2)} & k = 1; \sigma = \overline{1, n}, \end{cases}$$

$$d_{vk} = \frac{\prod_{\substack{j_0=1 \\ j_0 \neq k, j}}^{m} \left(t_{vj} - t_{vj_0}\right)}{\prod_{\substack{j_0=1 \\ j_0 \neq k}}^{m} \left(t_{vk} - t_{vj_0}\right)}.$$

In equation (2.1.1), we mean that $K(t_0, t), f(t_0) \in H_\alpha^r(L)$.

Let us prove that if equation (2.1.1) has a unique solution $\varphi_0 = \sqrt{\frac{t-a}{t-b}} \varphi(t)$, then starting from a certain $n = n_0$, equation (2.1.3') and system (2.1.3'') are uniquely solvable. Furthermore, if $L_n[\varphi_n; t_0]$ is a solution of equation (2.1.3'), then under the above assumptions regarding φ, the following estimate

$$\|\varphi(t) - L_n[\varphi_n; t]\|_{H_\beta} \leq \frac{C_r \ln n}{n^{r+\alpha-\beta-1/2}} \tag{2.1.4}$$

$$(m > r + 1, \beta < \alpha - 1/2)$$

holds true where φ is a solution to equation (2.1.3) and C_r is some constant. The validity of this theorem follows from the well-known results in the general approximation theory (see [38], Chapter XIV).

Let us note that based on the estimates obtained in §1.4, the following is valid

$$\|L_n[u[[L_n\varphi_n]] - U_n[L_n\varphi_n]\|_{H_\beta} \leq \frac{a_r \ln n}{n^{r+\alpha-\beta-\frac{1}{2}}} \|L_n\varphi_n\|_{H_\beta},$$

$$\left\| f_0 - L_n \left[\left(S_n^{(1/2;-1/2)} f \right); t_0 \right] \right\|_{H_\beta} \leq \frac{b_r \ln n}{n^{r+\alpha-\beta-\frac{1}{2}}} \| f_0 \|_{H_\beta},$$

where a_r, b_r are constants. As we can see, the estimates obtained in §1.4 are essential for the foundation of the given scheme.

It becomes clear from the results of the mentioned § that for the numerical solution of equation (2.1.1), the piece-wise interpolation sum $\left[L_n\left[\left(S_n^{(1/2;-1/2)} \varphi \right); t_0 \right] \right.$ constructed in §1.4 does not differ essentially from the sum $S_n^{(1/2;-1/2)}(\varphi; t_0)$ itself. Moreover, both sums generate the same system of linear equations. The sum $L_n \left[\left(S_n^{(1/2;-1/2)} \varphi \right); t_0 \right]$ considered in this § (and in the following ones as well) serves only as an auxiliary tool.

When $\alpha = 1$, from the estimate (2.1.4), according to the well-known theorem of Plemelj-Privalov (see [77] §18), under the corresponding assumptions on K and f, we can prove that the order of $\| \varphi - L_n \varphi \|$ is $O\left(\frac{1}{n^{r+1/2-\varepsilon}} \right)$, where ε is an arbitrarily small positive number.

§2.2 On the numerical solution of first kind singular integral equation, with open contours, in the case of a non-zero exponent

Let us again consider the equation

$$\frac{1}{\pi i} \int_L \frac{\varphi_0(t)dt}{t - t_0} + \frac{1}{\pi i} \int_L K(t_0, t)\varphi_0(t)dt = f(t_0) \qquad (t_0 \in L) \qquad (2.2.1)$$

and assume that $f(t), K(t_0, t) \in H$, we want to find a solution of equation (2.2.1), which is unbounded on both ends a and b.

As is known (see [77] §86), in this case, the index of equation (2.2.1) is equal to 1, and the solution, generally, is not uniquely determined.

Solving (2.2.1) with respect to the characteristic part (see [26] §42), it is reduced to an equivalent equation:

$$\begin{cases} \varphi(t_0) + \dfrac{1}{\pi i}\displaystyle\int_{ab} \dfrac{1}{\pi i}\displaystyle\int_{ab} \dfrac{K(t_1,t)dt}{\sqrt{(t-a)(t-b)}} \dfrac{dt_1}{t_1 - t_0}]\varphi_0(t)dt = \\ \displaystyle\int_{ab} \varphi_0(t)dt = C \end{cases}$$

$$= \dfrac{1}{\pi i}\int_{ab} \dfrac{f(t)dt}{\sqrt{(t-a)(t-b)}(t-t_0)}, \qquad\qquad (2.2.2)$$

where C is a constant.

In this case, the operator equation (2.1.3) will take the form:

$$V_\varphi = I_\varphi + k_1\varphi = f_1 \qquad\qquad (2.2.3)$$

where

$$\kappa_1\varphi = \dfrac{1}{\pi i}\int_{ab}\left[\dfrac{1}{\pi i}\int_{ab}\dfrac{K(t_1,t)}{\sqrt{(t-a)(t-b)}(t_1-t)}dt_1\right]\varphi_0(t)dt$$

$$f_1(t_0) \equiv \dfrac{1}{\pi i}\int_{ab}\dfrac{f(t)dt}{\sqrt{(t-a)(t-b)}(t-t_0)} + C$$

Similarly to the previous discussion, let us rewrite it as:

$$\dfrac{1}{\pi i}\int_{ab}\dfrac{K(t_1;t)}{\sqrt{(t-a)(t-b)(t-t_0)}}dt_1 \approx L_n\left[\left(S_n^{(-1/2;-1/2)}K\right);t,t_0\right], \quad t_0 \in ab$$

where

$$L_n\left[\left(S_n^{(-1/2;-1/2)}K\right);t,t_0\right] = L_{nj}\left[\left(S_n^{(-1/2;-1/2)}K\right);t,t_0\right],$$
$$t_0 \in \tau_j\tau_{j+1} \quad (j = \overline{1,n})$$

$$L_{nj}[\left(S_n^{(-1/2;-1/2)}K\right);\,t,t_0] = \sum_{k=1}^{m}\dfrac{\omega_j(t_0)}{(t_0 - t_{jk})\omega_j'(t_{jk})}S_n^{(-1/2;-1/2)}(K;t_{jk},t_0)$$
$$t_0 \in \tau_j\tau_{j+1}, \quad (j = \overline{1,n})$$

analogously

$$\frac{1}{\pi i}\int_{ab}\frac{1}{\sqrt{(t-a)(t-b)}}\frac{f(t)dt}{(t-t_0)} \approx L_n\left[\left(S_n^{(-1/2;-1/2)}f\right);t_0\right],$$

Together with equation (2.2.1), let us consider the equation

$$V_n[L_n[\varphi_n;t_0]] \equiv L_n[\varphi_n;t_0] + \frac{1}{\pi i}L_n\left[\left(S_n^{(1/2;1/2)}K\right);t_0t\right]L_n[\varphi_n;t_0]dt =$$

$$= L_n\left[\left(S_n^{(1/2;1/2)}f\right);t_0\right]$$

$$(2.2.1')$$

within a subclass of functions $L_n[\varphi_n;t_0]$. This equation, taking t_0 from set

$$T(v;j) = \{t_{vj}\} \qquad (v = 1,2,\dots,n; \ \ j = 1,2,\dots,m)$$

will bring us to linear system with unknowns $\varphi_n(t_{vj})$:

$$\begin{cases} \left[V_n[L_n[\varphi_n;t_0]]\right]_{t_0=t_{vj}} = \left[L_n\left[\left(S_n^{(-1/2;-1/2)}f\right);t_0\right]\right]_{t_0=t_{vj}} \\ (v = 1,2,\dots,n-1; \ \ j = 1,2,\dots,m) \\ \displaystyle\sum_{\sigma=1}^{n}\sum_{k=1}^{m} q_{\sigma k}^{*(1/2;1/2)}\varphi(t_{\sigma k}) = C; \quad v = n \end{cases} ,$$

or

$$\begin{cases} \displaystyle\varphi_n(t_{vj}) + \sum_{\sigma=1}^{n}\sum_{e=1}^{m} q_{ie}^{*(1/2;1/2)}B_n(t_{ie};t_{vj}) = f_{1n}(\iota_{vj}) \\ (v = 1,2,\dots,n-1; \ \ j = 1,2,\dots,m) \\ (v = n; \ \ j = 1,2,\dots,m-1) \\ \displaystyle\sum_{\sigma=1}^{n}\sum_{k=1}^{m} q_{\sigma k}^{*(1/2;1/2)}\varphi(t_{\sigma k}) = C; \end{cases} \qquad (2.2.3)$$

where

$$B_n(t_{ie},t_{vj}) = \left[\sum_{\substack{\sigma=1 \\ \sigma \neq v}}^{n}\sum_{k=1}^{m}\frac{P_{\sigma k}^{*(-1/2;-1/2)}}{t_{\sigma k}-t_{vj}} + \sum_{\substack{k=1 \\ k \neq j}}^{m}\frac{P_{vk}^{*(-1/2;-1/2)}}{t_{vk}-t_{vj}} - \right.$$

$$- P_{vj}^{*(-1/2;-1/2)} \sum_{\substack{k=1 \\ k \neq j}} d_{vk}(t_{vj}) \Bigg] K(t_{vj}, t_{ie}) -$$

$$- \sum_{\substack{\sigma=1 \\ \sigma \neq v}}^{n} \sum_{k=1}^{m} \frac{P_{\sigma k}^{*(-1/2;-1/2)}}{t_{\sigma k} - t_{vj}} K(t_{\sigma k}, t_{ie}) - \sum_{\substack{\sigma=1 \\ \sigma \neq j}}^{n} \frac{P_{vk}^{*(-1/2;-1/2)}}{t_{vk} - t_{vj}} K(t_{vk}, t_{ie}) +$$

$$+ P_{oj}^{*} \sum_{\substack{k=1 \\ k \neq j}}^{m} d_{vk}(t_{vj}) K(t_{vk}, t_{ie})$$

$\varphi_n(t_{vj})$ are again unknown functions.

$$f_{ln}(t_{vj}) = \Bigg[\sum_{\substack{\sigma=1 \\ \sigma \neq v}}^{n} \sum_{k=1}^{m} \frac{P_{\sigma k}^{*(-1/2;-1/2)}}{t_{vj} - t_{\sigma k}}$$

$$+ \sum_{\substack{k=1 \\ k \neq j}}^{m} \frac{P_{vk}^{*(-1/2;-1/2)}}{t_{vj} - t_{vk}} - P_{vj}^{*(-1/2;-1/2)} \sum_{\substack{k=1 \\ k \neq j}}^{m} d_{vk}(t_{vj}) \Bigg] \times$$

$$\times f(t_{vj}) - \sum_{\substack{\sigma=1 \\ \sigma \neq v}}^{n} \sum_{k=1}^{m} \frac{P_{\sigma k}^{*(-1/2;-1/2)}}{t_{vj} - t_{\sigma k}} f(t_{\sigma k}) - \sum_{\substack{k=1 \\ k \neq j}}^{m} \frac{P_{vk}^{*(-1/2;-1/2)}}{t_{vj} - t_{vk}} f(t_{vk}) +$$

$$+ P_{vj}^{*(-1/2;-1/2)} \sum_{\substack{k=1 \\ k \neq j}}^{m} d_{vk}(t_{vk}) f(t_{vk})$$

$$P_{\sigma k.}^{*(-1/2;-1/2)} = \frac{1}{\pi i} \int_{\tau_{\sigma} \tau_{\sigma+1}} \frac{1}{\sqrt{(t-a)(t-b)}} \prod_{\substack{j=1 \\ j \neq k}}^{m} \frac{t - t_{\sigma j}}{t_{\sigma k} - t_{\sigma j}} dt$$

$$q_{il}^{*(1/2;1/2)} = \frac{1}{\pi i} \int_{\tau_i \tau_{i+1}} \sqrt{(t-a)(-b)} \prod_{\substack{j=1 \\ j \neq e}}^{m} \frac{t - t_{ij}}{t_{ie} - t_{ij}}$$

$$P_{\sigma k}^{*(-1/2;-1/2)} = \begin{cases} P_{\sigma k}^{(-1/2;-1/2)}, & k = \overline{2, m-1}; \sigma = \overline{1, n}, \\ P_{\sigma 1}^{*(-1/2;-1/2)} + P_{\sigma-1,m}^{*(1/2;-1/2)} & k = 1; \sigma = \overline{1, n}, \end{cases}$$

$$d_{vk} = \frac{\prod_{\substack{j_0=1 \\ j_0 \neq k, j}}^{m} t_{vj} - t_{vj_0}}{\prod_{\substack{j=1 \\ j_0 \neq k}}^{m} t_{vk} - t_{vj_0}}$$

$$q_{\sigma k}^{*(1/2;1/2)} = \begin{cases} q_{\sigma k}^{(1/2;1/2)}, & k = \overline{2, m-1}; \sigma = \overline{1, n}, \\ q_{\sigma 1}^{(1/2;1/2)} + P_{\sigma-1m}^{(1/2;1/2)} & k = 1; \sigma = \overline{1, n}. \end{cases}$$

Let us again assume that $K, f \in H_\alpha^{(r)}(L)$. It can be proven that if equation (2.2.1) has a unique solution (in this case, it is assumed that the integral from the sought solution is known) $\int_{ab} \varphi_0(t)dt = C$) $\quad \varphi_0(t) = \frac{\varphi(t)}{\sqrt{(t-a)(t-b)}}$, then starting from a certain $n = n_0$, equation (2.2.1') and system (2.2.3) have a unique solution. Moreover, if $L_n(\varphi_n; t_o)$ is the solution of equation (2.2.1'), then under the above assumptions for K and f, the following estimates hold

$$\|\varphi(t) - L_n[\varphi_n, t]\|_{H_\beta} \leq \frac{A_r \ln n}{n^{r+\alpha-\beta-\frac{1}{2}}}, \quad (m > r+1, \beta < \alpha - 1/2), \quad (2.2.4)$$

where A_r is a constant that depends only on L and the Hölder exponent of the function $\varphi^{(r)}$.

The validity of this theorem follows from the general theory of approximation (see [38], Chapter XIV) and the estimates obtained by us in §1.4

$$\left\|L_n[V[L_n\varphi_n]] - V[L_n\varphi_n]\right\|_{H_\beta} \leq \frac{b' \ln n}{n^{r+\alpha-\beta-\frac{1}{2}}} \|L_n\varphi_n\|_{H_\beta},$$

$$\left\|f_1 - L_n\left[\left(S_n^{-1/2;-1/2}f\right); t_0\right]\right\|_{H_\beta} \leq \frac{a_r' \ln n}{n^{r+\alpha-\beta-\frac{1}{2}}} \|f_1\|_{H_\beta},$$

where b' and a' are constants. As we can see, the estimates from §1.4 also play an essential role in the foundation of this scheme.

R e m a r k : In the case when the index of equation (2.2.1) is negative $\aleph = -1$, there does not exist always an exact solution to the given equation. For this solution to exist, the right-hand side $f(t_0)$ must satisfy an additional condition

$$\int_{ab} \frac{f(t)dt}{\sqrt{(t-a)(t-b)}} = 0 \qquad (2.2.5)$$

Therefore, when solving equation (2.2.1) approximately, the additional condition (2.2.5) must be taken into account. As is known (see, e.g., [77]), if the condition (2.2.5) is satisfied, then equation (2.2.1) always has a solution, and this solution is unique.

Now, let us solve the first-kind singular integral equation of the following form using the above-mentioned algorithm, considering different values of the parameters involved (We will examine this equation in greater detail in Chapter V, where we will discuss the numerical solutions of practical problems of an applied type.

Table 7 $n = 3$; $m = 2$; $H = 1$; $f = 2$ (the exact solution is $\varphi = 2$)

t_{vj}	-1	-0.3333	0.3333	1
$\varphi(t_{vj})$	1.85	1.98	2.08	2.13

Table 8 $n = 3$; $m = 4$; $H = 1$; $f = 2$

t_{vj}	-1	-0.778	-556	-0.333	-0.111	0.111	0.333	0.556	0.778	1
$\varphi(t_{vj})$	$\frac{1.8}{2}$	1.86	1.90	1.94	1.98	2.02	2.06	2.09	2.12	2.14

Table 9 $n = 5$; $m = 2$; $H = 1$; $f = 2$

t_{vj}	-1	-0.6	-0.2	-0.2	-0.6	1
$\varphi(t_{vj})$	1.82	1.902	1.981	0.050	2.104	2.138

§2.3 Numerical solution of the characteristic equation using the discrete singularities method with an increased order of accuracy

The intense spread of personal computers and its wide application have essentially united fundamental mathematical and applied problems.

Arising of the recent, significantly powerful numerical experiments in general research methods, as never before, has closely related the physical context of the problem to its mathematical formulation and numerical methods for its solution, counting peculiarities of personal computing technologies.

Describing the phenomenon naturally emerged when the entire process is studied comprehensively and its development is considered, a non-stationary and discrete approach became leading.

The increasing demands of practice lead to the complexity of applied problems, for which traditional numerical methods often prove to be ineffective.

Often, a researcher of applied problems quickly identifies a rational solution path, as it relies on understanding the fundamental essence of the problem.

In such an approach, numerical experiment helps us to identify a rational progression of the search, after which a situation arises where mathematical foundations and generalizations can be established.

For the success of this process, it becomes crucial to develop methods that integrate the three main aspects of problem: physical, mathematical, and computational.

It should be emphasized that the role of mathematics is not limited to foundations alone; rather, neither by generalization of a method, but, it and only it is capable to make a method universal that is not confined to only solving problems of a specific type. In this regard, the method of discrete singularities is of particular interest.

In the process of constructing approximations for singular integrals by this scheme, the effect is enhanced by the fact that for singular integral equations, it is possible to directly inverse the system of linear algebraic equations, along with estimation of the corresponding increment of the method.

A significant interest arises when, inspite of the fact that the integrand is being approximated with entirely unbounded kernel, position of so called computational and control points give us possibility to prove convergence of the method. Additionally, it becomes possible to evaluate the convergence order.

Under the assumption that the density of the singular integral satisfies the Hölder condition, in most cases, the estimate $O\left(\frac{\ln n}{n^{\alpha}}\right)$, $(0 \leq \alpha \leq 1)$ holds true.

In the case when the integration line is a real segment, the nodes are defined by the roots of Jacobi polynomial [66], (with corresponding indeces) and the control nodes represent null points linked to them [66] (see also [43], [82]), the method of discrete singularitites gives Gauss accuracy of approximation of the corresponding integrals.

But, in general, using dependence of equations and intregrals on line, we cannot prove accuracy order higher than $O(\ln n/n)$. It is reasonable to question whether it is possible to modify the mentioned schemes in order to keep in a certain sense those positive properties they already have. In addition, it is needed to increase their convergence rate in the case of any smooth contour. Below, the mentioned question is discussed.

Let us consider a singular integral equation of the first kind

$$\frac{1}{\pi i}\int_L \frac{\varphi(t)dt}{t - t_0} + \frac{1}{\pi i}\int_L K(t_0, t)\varphi(t)dt = f(t_0), \qquad (2.3.0)$$

where $L = ab$ is an open smooth contour with endpoints a and b given parametrically $t = t(s)(s_a \leq s \leq s_b)$.

Let us also assume that the function f(t0) possesses a certain smoothness, while the kernel $K(t_0, t)$ may have an integrable diagonal singularity. To be more specific $K(t_0, t) = h(t_0, t)/(t - t_0)$.At this, vector-function $\left(h(t_0, t),\ f(t_0)\right) \in H_a$ (satisfies the Hölder condition on L with exponent $0 < \alpha \leq 1$).

As mentioned, equations of the form (2.3.0) correspond to many applied problems, including those in structural mechanics (see, for example, [82]). It is known that, for solving singular integral equations of this type and others numerically, so-called direct computational methods are of particular interest.

Below, a computational scheme is constructed, which, similar to the classical discrete singularities method, leads us to a process that is relatively easy to implement. Additionally, these schemes are based on the direct division of the contour L and do not require prior mapping of this contour onto a segment.

In [99], for the numerical solution of equation (2.3.0) (when the index of this equation (see [77]) is equal to zero), where L is a closed Lyapunov contour, a direct computational algorithm is constructed, which improves the classical discrete singularity method in terms of accuracy.

Below, the construction of a similar direct algorithm is presented, along with corresponding estimates, for the case where $L = ab$ is an open smooth contour.

In this case, because the singular integral is generally unbounded at the contour's endpoints (see [77]), additional significant difficulties arise.

Let's develop the characteristic first -kind singular integral equation

$$\frac{1}{\pi i}\int_L \frac{\varphi(t)dt}{t - t_0} = f(t_0). \tag{2.3.1}$$

As is known from [77], depending on the class of functions we are seeking the solution in, for the integral at one end being bounded, at the other end unbounded (index $\aleph = 0$), unbounded at both ends ($\aleph = 1$), or bounded at both ends ($\aleph = -1$), the exact solutions of the equation (2.3.1) take different forms.

When $\aleph = 1$, all solutions of the equation (2.3.1) (of the class $h(1)$) are expressed in the form

$$\varphi(t_0) = \frac{1}{\pi i\sqrt{(t_0 - a)(t_0 - b)}}\int_L \frac{\sqrt{(t - a)(t - b)}f(t)dt}{t - t_0} +$$

$$+ \frac{C}{\sqrt{(t_0 - a)(t_0 - b)}}, \tag{2.3.2}$$

where C is an arbitrary constant and

$$\int_L \varphi(t)dt = C; \tag{2.3.3}$$

When $\aleph = 0$, it has a unique solution (of class $h(1)$) of the following form:

$$\varphi(t_0) = \frac{1}{\pi i}\sqrt{\frac{t_0 - b}{t_0 - a}}\int_L \sqrt{\frac{t - a}{t - b}}\frac{f(t)dt}{t - t_0}, \tag{2.3.4}$$

and a unique solution (of the class $h(-1)$) of the following form

$$\varphi(t_0) = \frac{1}{\pi i}\sqrt{\frac{t_0 - a}{t_0 - b}}\int_L \sqrt{\frac{t - b}{t - a}}\frac{f(t)dt}{t - t_0}, \tag{2.3.5}$$

When $\aleph = -1$ then it has a unique solution (of the class $h(-1,1)$)

$$\varphi(t_0) = \frac{\sqrt{(t_0 - a)(t_0 - b)}}{\pi i} \int_L \frac{f(t)dt}{\sqrt{(t - a)(t - b)}}, \qquad (2.3.6)$$

under additional condition

$$\int_L \frac{f(t)dt}{\sqrt{(t - a)(t - b)}} = 0. \qquad (2.3.7)$$

Now, let's construct an approximate (numerical) solution algorithm for the equation (2.3.1).

Let's assume that we are looking for a solution to equation (2.3.1), which is bounded at the endpoint a and is unbounded at b. As we have noted, it has the following form:

$$\varphi_0(t) = \sqrt{\frac{t - a}{t - b}}\, \varphi(t). \qquad (2.3.8)$$

Let's substitute (2.3.8) into (2.3.1) and take into account that $\frac{1}{\pi i} \int_L \sqrt{\frac{t-a}{t-b}} \frac{dt}{t-t_0} = 1$, then equation (2.3.1) can be transformed into its equivalent equation

$$\varphi(t_0) + \frac{1}{\pi i} \int_{ab} \sqrt{\frac{t - a}{t - b}} \frac{\varphi(t) - \varphi(t_0)}{t - t_0} dt = f(t_0). \qquad (2.3.9)$$

Generally, foundation of approximate solution process on an arbitrary smooth open contour for this equation is complex, but computer-based research allows us to choose the system of nodes $\{t_{0j}\}$ in such a way that the approximate system of equations will have a solution tending to the solution of exact equation with a certain rate.

Let us change the expression

$$\frac{1}{\pi i} \int_{ab} \sqrt{\frac{t - a}{t - b}} \frac{\varphi(t) - \varphi(t_0)}{t - t_0} dt$$

in equation (2.3.9) by the quadrature formula $D_n^{(1/2;-1/2)}\left[(S^{(-1/2;-1/2)}\varphi); t_0\right]$, constructed in §1.5, and choose the roots of the following polynomial $p_1 \prod_{j=2}^{n+1}(t_j - t) + \sum_{\sigma=1}^{n-1}(p_{\sigma+1} + p1_\sigma)\left(\prod_{j=1}^{n+1}(\tau_j - t)\right)\left[(\tau_{\sigma+1} - t) + p1_n \prod_{j=1}^{n}(\tau_j - t)\right] -$

$$-\prod_{j=1}^{n+1}(\tau_j - t) = 0. \tag{2.3.10}$$

as discrete points $\{t_{0j}\}$ corresponding to t_0. As computer-based research showed, the roots of the given polynomial are located on the contour $L = ab$. At this, they almost coinside with points $t_{0j} = (\tau_j + \tau_{j+1})/2$, $(j = 1,2,\dots,2n)$.

Then, since the given polynomial is derived from the quadrature formula used to approximately compute $\frac{1}{\pi i}\int_{ab}\sqrt{\frac{t-a}{t-b}}\frac{\varphi(t)}{t-t_0}dt$, if we take the mentioned roots of the polynomial as the values of $\{t_{0j}\}$, the norm of the approximate operator constructed by us will be close to the norm of the corresponding operator in the classical discrete singularity method.

Therefore, starting from a specific n_0 of division of contour L=ab, our system's solution will also exist, and we can write the corresponding estimations.

Let us fix $v_0(1 \le v_0 \le 2n)$ and construct on L a piecewise continuous quadratic interpolation polynomial $(G_n\varphi)(t_0)$ using the values of the sum $\sum_{\sigma=1}^{n}$ at three consecutive nodes $\tau_{2v_0-1}, \tau_{2v}, \tau_{2v+1}$.

Assuming $t_0 \in \tau_{2v_0-1}\tau_{2v_0+1}$, $(v_0 = 1,2,\dots.n)$, we obtain an approximation of the corresexpression ofar integral for any $t_0 \in L$ in the form of an expression of the type $\varphi(t_0) + (G_n\varphi)(t_0)$.

In the space (subspace) of piecewise continuous polynomial functions w_n on L, we consider

$(D_n w_n)(t_0) \equiv w_n(t_0) + (G_n w_n)(t_0), t_0 \in L.$

Let us demonstrate that for sufficiently large n, the operators D_n are continuously invertible in the corresponding subspace. To analyze the problem, consider $(G_n w_n)(\tau_{2v})$, where the odd-indexed points of the discretization are taken as the nodes.

$$(G_n w_n)(\tau_{2v}) = w(\tau_{2v}) + p_1 \chi(\tau_1, \tau_{2v}) +$$

$$+ \sum_{\sigma=1}^{n-1} (p1_{2\sigma-1} + p_{2\sigma+1}) \chi(\tau_{2\sigma+1}, \tau_{2v}) + p1_{2n-1}\chi(\tau_{2n+1}, \tau_{2v}) =$$

$$= w(\tau_{2v}) + p_1 \frac{w(\tau_1) - w(\tau_{2v})}{\tau_1 - \tau_{2v}} +$$

$$+ \sum_{\sigma=1}^{n-1} (p1_{2\sigma-1} + p_{2\sigma+1}) \frac{w(\tau_{2\sigma+1}) - w(\tau_{2v})}{\tau_{2\sigma+1} - \tau_{2v}} + p1_{2n-1} \frac{w(\tau_{2n+1}) - w(\tau_{2v})}{\tau_{2n+1} - \tau_{2v}} =$$

$$= w(\tau_{2v}) + p_1 \frac{w(\tau_1)}{\tau_1 - \tau_{2v}} + \sum_{\sigma=1}^{n-1} (p1_{2\sigma-1} + p_{2\sigma+1}) \frac{w(\tau_{2\sigma+1})}{\tau_{2\sigma+1} - \tau_{2v}}$$

$$+ p1_{2n-1} \frac{w(\tau_{2n+1})}{\tau_{2n+1} - \tau_{2v}} -$$

$$- w(\tau_{2v}) \left(\frac{p_1}{\tau_1 - \tau_{2v}} + + \sum_{\sigma=1}^{n-1} \frac{(p1_{2\sigma-1} + p_{2\sigma+1}) \dfrac{1}{\tau_{2\sigma+1} - \tau_{2v}} + p1_{2n-1}}{\dfrac{1}{\tau_{2n+1} - \tau_{2v}}} \right)$$

$$(2.3.11)$$

Let us select the sum $\sum 2v$ from (2.3.11) with factor $w_n(\tau_{2v})$:

$$\sum 2v = - w_n(\tau_{2v}) \Big(\frac{p_1}{\tau_1 - \tau_{2v}}$$

$$+ \sum_{\sigma=1}^{n-1} (p1_{2\sigma-1} + p_{2\sigma+1}) \frac{1}{\tau_{2\sigma+1} - \tau_{2v}} + p1_{2n-1} \frac{1}{\tau_{2n+1} - \tau_{2v}} \Big) =$$

$$= w_n(\tau_{2v}) \{ p_1 \prod_{j=2}^{n+1} (\tau_{2j-1} - \tau_{2v}) + \sum_{\sigma=1}^{n-1} (p1_{2\sigma-1} + p_{2\sigma+1}) \Big(\prod_{j=1}^{n+1} \frac{\tau_{2j-1} - \tau_{2v}}{\tau_{2\sigma+1} - \tau_{2v}} +$$

$$+ p1_{2n-1} \frac{\prod_{j=1}^{n}(\tau_{2j} - \tau_{2v})}{\prod_{j=1}^{n+1}(\tau_{2j} - \tau_{2v})} \}.$$

If we consider that the nodal points $\{t_{2v}\}$ are the roots of the polinomial (2.3.10), then it becomes evident that $\sum 2v \to 1$, and consequently, we denote

$$(\tilde{G}_n w_n)(\tau_{2v}) = w_n(\tau_1) \frac{p_1}{\tau_1 - \tau_v} +$$

$$+ \sum_{\sigma=1}^{n-1} (p1_{2\sigma-1} + p_{2\sigma+1}) \frac{w_n(\tau_{2\sigma+1})}{\tau_{2\sigma+1} - \tau_{2v}}$$

$$+ p1_{2n-1} \frac{w_n(\tau_{2n+1})}{\tau_{2n+1} - \tau_{2v}}, \qquad (2.3.12)$$

then for any $v(1 \leq v \leq n)$

$$(G_n w_n)(\tau_{2v}) = (\tilde{G}_n w_n)(\tau_{2v}) - \sum nv w_n(\tau_{2v}),$$

where $\sum nv \to 0, (n \to \infty)$, consequently, under $(\tilde{G}_n w_n)(t_0)$ we mean piecewise interpolation polynomial constructed by values of $(\tilde{G}_n w_n)(\tau_{2v})$. Therefore we can write

$$(D_n w_n)(t_0) = (G_n w_n)(t_0) + \varepsilon_n(w_n; t_0), \qquad (2.3.13)$$

where

$$\max_{t_0} |\varepsilon_n(w_n; t_0)| \leq \varepsilon_n \max_t |w_n(t)|, \qquad \varepsilon_n \to 0, (n \to \infty).$$

For the given n, in the corresponding subspace, we consider homogeneous equation

$$(\tilde{G}_n w_n)(t_0) = 0. \qquad (2.3.14)$$

If we now construct the expression of $(\tilde{G}_n w_n)(t_0) = 0$ so that the control points (in our case, the even-indexed division points), are taken as nodal points, and nodal points (the odd-indexed division points) are taken as control points, then the representation of $(\tilde{G}_n w_n)(t_0) = 0$ at th points $\{\tau_{2j-1}\}(j = 1, 2, \dots n + 1)$ will have the following form:

$$(\tilde{G}_n w_n)(\tau_{2j-1}) = w_n(\tau_2) \frac{p_2}{\tau_2 - \tau_{2j-1}} +$$

$$+ \sum_{\sigma=1}^{n-1} (p1_{2\sigma} + p_{2\sigma+2}) \frac{w_n(\tau_{2\sigma+2})}{\tau_{2\sigma+2} - \tau_{2j-1}} + p1_{2n-2} \frac{w_n(\tau_{2n})}{\tau_{2n} - \tau_{2j-1}} \qquad (2.3.15)$$

Let us take into account the structures of the expressions (2.3.12) and (2.3.15). We observe that (2.3.14) represents two independent linear systems of equations.

From the structure of the coefficients $p1_j, p_j (j = 1, 2, \dots 2n)$ (see §1.5), it is evident that for a smooth line L, all of them differ from zero.

Let us consider the above mentioned and the known results (see [16], [66]). Regarding the structure of the coefficients in systems (2.3.12) and (2.3.15), it can be concluded that for both systems, it is possible to employ the classical discrete singularities method.

Based on the above, it can be concluded that equation (2.3.14) has only the trivial solution, from which it follows, in its turn, that the operator $\tilde{G}_n$ is continuously invertible. From this, based on (2.3.13), we conclude that for sufficiently large n, the operators D_n are also continuously invertible.

Let's use the well-known approach to the study of inverse matrices in discrete singularities method (see [16], [66]).

It can be concluded that the norms of the inverse operators D_n^{-1} (inverse of D_n) in the corresponding space have an estimate $O(n^{1/2}\ln n), (n \to \infty)$.

Based on this, it can be noted that the corresponding inverse operators can be defined on the class of Hölder functions, and it can be demonstrated that their norms are estimated by the corresponding exponent for these classes.

Next, based on this reasoning, we will discuss the Hölder space H_β with the corresponding exponent $\beta < \alpha - 1/2, (1/2 < \alpha \le 1)$. In addition, for the corresponding norm, let us use the notation $\|D_n^{-1}\|_{H_\beta}$.

If we take into account the above remark regarding the roots of the polynomial (2.3.10), we can be convinced of the fairness of the estimate $\|D_n^{-1}\|_{H_\beta} = O(n^{1/2+\beta}\ln^2 n , (n \to \infty)$.

Such a conclusion is primarily based on estimation of expression

$$\sup_{t_0, t \in L_1} \frac{|(G_n\varphi)(t_0) - (G_n\varphi)(t_1)|}{|t_0 - t_1|^\beta}. \tag{2.3.16}$$

By $\max_j |\varphi(\tau_j)|$.

Let us assume from the beginning that $|t_0 - t_1| > \min_j |t_{j+1} - t_j|$. Evidently,

$$\frac{|(G_n\varphi)(t_0) - (G_n\varphi)(t_1)|}{|t_0 - t_1|^\beta} \leq \frac{2\max_j |(G_n\varphi)(t_j)|}{\min_j |t_{j+1} - t_j|^\beta}.$$

Let's take into account smoothness of the line $L = ab$ and $\max_i |p_i| = O(n^{-1/2})$. For $\max_j |(G_n\varphi)(t_j)|$ we easily obtain the estimation, which for the expression under consideration leads us to estimation $O(n^{1/2+\beta}\ln n)\max_i|\varphi(t_i)|$.

Now let's consider the values of t_0, t_1 that correspond to the condition $|t_0 - t_1| \leq \max_j |t_{j+1} - t_j|$.

This case clearly leads to the situation where, in the evaluation of (2.3.16), the corresponding points t_0, t_1 are located on the same arc of the division of the line $L = ab$.

$$|(G_n\varphi)(t_0) - (G_n\varphi)(t_1)| \leq c_1 n |t_0 - t_1| \max_\nu |(G_n\varphi)(\tau_\nu)|.$$

In this case, if we use the structure of Lagrange interpolation coefficients for the expression $(G_n\varphi)(t_0)$, we can easily obtain the inequality

$$|(G_n\varphi)(t_0) - (G_n\varphi)(t_1)| \leq c_1 n |t_0 - t_1| \max_\nu |(G_n\varphi)(\tau_\nu)|.$$

from which the needed estimation follows easily. The assessment of the order of convergence of the scheme in our assumption can be obtained based on the estimation of the residual term of the corresponding approximation scheme $D_n^{(1/2;-1/2)}(\varphi; t_0)$ of the integral $S^{(1/2;-1/2)}(\varphi; t_0)$ in the space of Hölder functions.

Similarly, the reasoning applies to the corresponding characteristic equations for $S^{(-1/2;-1/2)}(\varphi; t_0)$ and $S^{(1/2;1/2)}(\varphi; t_0)$.

Let's write the corresponding systems in expanded form:

$$\left[1 - \frac{p1_1^{(1/2;-1/2)}}{\tau_1 - \tau_{2j}} - \sum_{\sigma=1}^{n-1}\frac{\left(p1_{2\sigma-1}^{(1/2;-1/2)} + p_{2\sigma+1}^{(1/2;-1/2)}\right)}{\tau_{2\sigma+1} - \tau_{2j}} - \frac{p1_{2n-1}^{(1/2;-1/2)}}{\tau_{2n+1} - \tau_{2j}}\right]\varphi(\tau_{2j}) +$$

$$+\frac{p_1^{(1/2;-1/2)}}{\tau_1-\tau_{2j}}\,\varphi(\tau_1)+$$

$$+\left[\sum_{\sigma=1}^{n-1}\frac{\left(p1_{2\sigma-1}^{(1/2;-1/2)}+p_{2\sigma+1}^{(1/2;-1/2)}\right)}{\tau_{2\sigma+1}-\tau_{2j}}\,\varphi(\tau_{2\sigma+1})+\frac{p1_{2n-1}^{(1/2;-1/2)}}{\tau_{2n+1}-\tau_{2j}}\,\varphi(\tau_{2\sigma})\right]$$

$$=f(\tau_{2j})$$

$$j=1,2,\ldots n.$$

$$\left[1-\frac{p_2^{(1/2;-1/2)}}{\tau_2-\tau_{2j-1}}-\sum_{\sigma=1}^{n-1}\frac{\left(p1_{2\sigma}^{(1/2;-1/2)}+p_{2\sigma+2}^{(1/2;-1/2)}\right)}{\tau_{2\sigma+2}-\tau_{2j-1}}-\frac{p1_{2n-2}^{(1/2;-1/2)}}{\tau_{2n}-\tau_{2j-1}}\right]\varphi(\tau_{2j-1})$$

$$+\frac{p_2^{(1/2;-1/2)}}{\tau_2-\tau_{2j-1}}\,\varphi(\tau_2)+$$

$$+\left[\sum_{\sigma=1}^{n-1}\frac{\left(p1_{2\sigma}^{(1/2;-1/2)}+p_{2\sigma+2}^{(1/2;-1/2)}\right)}{\tau_{2\sigma+2}-\tau_{2j-1}}\,\varphi(\tau_{2\sigma+2})+\frac{p1_{2n-2}^{(1/2;-1/2)}}{\tau_{2n}-\tau_{2j-1}}\,\varphi(\tau_{2\sigma})\right]$$

$$=f(\tau_{2j-1}),$$

$$j=1,2,\ldots n+1,$$

$$(2.3.17)$$

where

$$p1_{2\sigma-1}^{(1/2;-1/2)}=\frac{1}{\pi i}\int_{\tau_{2\sigma-1}\tau_{2\sigma+1}}\sqrt{\frac{t-a}{t-b}}\,\frac{(\tau_{2\sigma+1}-t)}{(\tau_{2\sigma+1}-\tau_{2\sigma-1})}\,dt$$

$$p_{2\sigma+1}^{(1/2;-1/2)}=\frac{1}{\pi i}\int_{\tau_{2\sigma-1}\tau_{2\sigma+1}}\sqrt{\frac{t-a}{t-b}}\,\frac{t-\tau_{2\sigma-1}}{\tau_{2\sigma+1}-\tau_{2\sigma-1}}\,dt$$

$$p1_{2\sigma.}^{(1/2;-1/2)}=\frac{1}{\pi i}\int_{\tau_{2\sigma}\tau_{2\sigma+2}}\sqrt{\frac{t-a}{t-b}}\,\frac{(\tau_{2\sigma+2}-t)}{(\tau_{2\sigma+2}-\tau_{2\sigma})}\,dt$$

$$p_{2\sigma+2}^{(1/2;-1/2)}=\frac{1}{\pi i}\int_{\tau_{2\sigma}\tau_{2\sigma+2}}\sqrt{\frac{t-a}{t-b}}\,\frac{t-\tau_{2\sigma}}{\tau_{2\sigma+2}-\tau_{2\sigma}}\,dt$$

$$(\sigma=1,2,\ldots n-1).$$

If we are looking for an unbounded (at both ends) solution of the characteristic equation, then the roots of the following polynomial

$$p_1 \prod_{j=2}^{n+1}(t_j - t)$$

$$+ \sum_{\sigma=1}^{n-1}(p_{\sigma+1} + p1_\sigma)(\prod_{j=1}^{n+1}(\tau_j - t))((\tau_{\sigma+1} - t) + p1_n \prod_{j=1}^{n}(\tau_j - t)]$$

$$= 0,$$

should be taken as calculation nodes $\{t_{0j}\}$, where coefficients $\{p_j\}, \{p1\}$ are calculated by formula (2.3.19). The corresponding system of linear equations regarding the unknowns $\varphi(\tau_j)$ $(j = 1,2, \dots 2n + 1)$ will have the form

$$\left[\frac{p_1^{(-1/2;-1/2)}}{\tau_1 - \tau_{2j}} + \sum_{\sigma=1}^{n-1}\frac{\left(p1_{2\sigma-1}^{(-1/2;-1/2)} + p_{2\sigma+1}^{(-1/2;-1/2)}\right)}{\tau_{2\sigma+1} - \tau_{2j}} + \frac{p1_{2n-1}^{(-1/2;-1/2)}}{\tau_{2n+1} - \tau_{2j}}\right]\varphi(\tau_{2j}) +$$

$$+ \frac{p_1^{(-1/2;-1/2)}}{\tau_1 - \tau_{2j}}\varphi(\tau_1) +$$

$$+ \left[\sum_{\sigma=1}^{n-1}\frac{\left(p1_{2\sigma-1}^{(-1/2;-1/2)} + p_{2\sigma+1}^{(-1/2;-1/2)}\right)}{\tau_{2\sigma+1} - \tau_{2j}}\varphi(\tau_{2\sigma+1}) + \frac{p1_{2n-1}^{(-1/2;-1/2)}}{\tau_{2n+1} - \tau_{2j}}\varphi(\tau_{2\sigma})\right]$$

$$= f(\tau_{2j}),$$
$$j = 1,2, \dots n$$

$$(2.3.18)$$

$$\left[\frac{p_2^{(-1/2;-1/2)}}{\tau_2 - \tau_{2j-1}} + \sum_{\sigma=1}^{n-1}\frac{\left(p1_{2\sigma}^{(-1/2;-1/2)} + p_{2\sigma+2}^{(-1/2;-1/2)}\right)}{\tau_{2\sigma+2} - \tau_{2j-1}} + \frac{p1_{2n-2}^{(-1/2;-1/2)}}{\tau_{2n} - \tau_{2j-1}}\right]\varphi(\tau_{2j-1})$$

$$+$$

$$+ \frac{p_2^{(-1/2;-1/2)}}{\tau_2 - \tau_{2j-1}}\varphi(\tau_2) +$$

$$+ \left[\sum_{\sigma=1}^{n-1}\frac{\left(p1_{2\sigma}^{(-1/2;-1/2)} + p_{2\sigma+2}^{(-1/2;-1/2)}\right)}{\tau_{2\sigma+2} - \tau_{2j-1}}\varphi(\tau_{2\sigma+2}) + \frac{p1_{2n-2}^{(-1/2;-1/2)}}{\tau_{2n} - \tau_{2j-1}}\varphi(\tau_{2\sigma})\right] =$$

$$= f(\tau_{2j-1}), \qquad j = 1,2, \dots n$$
$$p_2^{(-1/2;-1/2)}\varphi(\tau_{2\sigma}) +$$

$$+ \sum_{\sigma=1}^{n-1}\left(p1_{2\sigma}^{(-1/2;-1/2)} + p_{2\sigma+2}^{(-1/2;-1/2)}\right)\varphi(\tau_{2\sigma+2}) + p1_{2n-2}^{(-1/2;-1/2)}\varphi(\tau_{2n}) = C,$$

where C is some constant and the system coefficients can be presented as

$$p1_{2\sigma-1}^{(-1/2;-1/2)} = \frac{1}{\pi i} \int_{\tau_{2\sigma-1}\tau_{2\sigma+1}} \frac{1}{\sqrt{(t-a)(t-b)}} \frac{(\tau_{2\sigma+1}-t)}{(\tau_{2\sigma+1}-\tau_{2\sigma-1})} dt$$

$$p_{2\sigma+1}^{(-1/2;-1/2)} = \frac{1}{\pi i} \int_{\tau_{2\sigma-1}\tau_{2\sigma+1}} \frac{1}{\sqrt{(t-a)(t-b)}} \frac{t-\tau_{2\sigma-1}}{\tau_{2\sigma+1}-\tau_{2\sigma-1}} dt$$

$$p1_{2\sigma}^{(-1/2;-1/2)} = \frac{1}{\pi i} \int_{\tau_{2\sigma}\tau_{2\sigma+2}} \frac{1}{\sqrt{(t-a)(t-b)}} \frac{(\tau_{2\sigma+2}-t)}{(\tau_{2\sigma+2}-\tau_{2\sigma})} dt \qquad (2.3.19)$$

$$p_{2\sigma+2}^{(-1/2;-1/2)} = \frac{1}{\pi i} \int_{\tau_{2\sigma}\tau_{2\sigma+2}} \frac{1}{\sqrt{(t-a)(t-b)}} \frac{t-\tau_{2\sigma}}{\tau_{2\sigma+2}-\tau_{2\sigma}} dt.$$

When we seek a solution bounded at either end, the edditional condition must be fulfilled

$$\int_L \frac{f(t)dt}{\sqrt{(t-a)(t-b)}} = 0.$$

In this case points $t_{oj} = \frac{\tau_j+\tau_{j+1}}{2}$ can be taken as nodes $\{t_{oj}\}$. The corresponding system of linear equations with unknowns $\varphi(\tau_j)$ $(j = 1,2,\ldots 2n+1)$ will have the form:

$$\gamma_{on} + \left[\frac{p_1^{(1/2;1/2)}}{\tau_1-\tau_{2j}} + \sum_{\sigma=1}^{n-1} \frac{\left(p1_{2\sigma-1}^{(1/2;1/2)}+p_{2\sigma+1}^{(1/2;1/2)}\right)}{\tau_{2\sigma+1}-\tau_{2j}} + \frac{p1_{2n-1}^{(1/2;1/2)}}{\tau_{2n+1}-\tau_{2j}} \right] \varphi(\tau_{2j}) +$$

$$\frac{p_1^{(1/2;1/2)}}{\tau_1-\tau_{2j}} \varphi(\tau_1) +$$

$$+ \left[\sum_{\sigma=1}^{n-1} \frac{\left(p1_{2\sigma-1}^{(1/2;1/2)} + p_{2\sigma+1}^{(1/2;1/2)}\right)}{\tau_{2\sigma+1} - \tau_{2j}} \varphi(\tau_{2\sigma+1}) + \frac{p1_{2n-1}^{(1/2;1/2)}}{\tau_{2n+1} - \tau_{2j}} \varphi(\tau_{2\sigma}) \right] = f(\tau_{2j})$$

$$j = 1,2,\ldots n$$

$$\gamma_{on} + \left[\frac{p_2^{(1/2;1/2)}}{\tau_2 - \tau_{2j-1}} + \sum_{\sigma=1}^{n-1} \frac{\left(p1_{2\sigma}^{(1/2;1/2)} + p_{2\sigma+2}^{(1/2;1/2)}\right)}{\tau_{2\sigma+2} - \tau_{2j-1}} + \frac{p1_{2n-2}^{(1/2;1/2)}}{\tau_{2n} - \tau_{2j-1}} \right] \varphi(\tau_{2j-1}) +$$

$$+ \frac{p_2^{(1/2;1/2)}}{\tau_2 - \tau_{2j-1}} \varphi(\tau_2) +$$

$$+ \left[\sum_{\sigma=1}^{n-1} \frac{\left(p1_{2\sigma}^{(1/2;1/2)} + p_{2\sigma+2}^{(1/2;1/2)}\right)}{\tau_{2\sigma+2} - \tau_{2j-1}} \varphi(\tau_{2\sigma+2}) + \frac{p1_{2n-2}^{(1/2;1/2)}}{\tau_{2n} - \tau_{2j-1}} \varphi(\tau_{2n}) \right] = f(\tau_{2j-1}),$$

$$j = 1,2,\ldots n+1$$

Where γ_{0n} is the correcting parameter [16], while the coefficients of the system are defined as follows:

$$p1_{2\sigma-1}^{(1/2;1/2)} = \frac{1}{\pi i} \int_{\tau_{2\sigma-1}\tau_{2\sigma+1}} \sqrt{(t-a)(t-b)} \frac{(\tau_{2\sigma+1}-t)}{(\tau_{2\sigma+1}-\tau_{2\sigma-1})} dt$$

$$p_{2\sigma+1}^{(1/2;1/2)} = \frac{1}{\pi i} \int_{\tau_{2\sigma-1}\tau_{2\sigma+1}} \sqrt{(t-a)(t-b)} \frac{t-\tau_{2\sigma-1}}{\tau_{2\sigma+1}-\tau_{2\sigma-1}} dt$$

$$p1_{2\sigma}^{(1/2;1/2)} = \frac{1}{\pi i} \int_{\tau_{2\sigma}\tau_{2\sigma+2}} \sqrt{(t-a)(t-b)} \frac{(\tau_{2\sigma+2}-t)}{(\tau_{2\sigma+2}-\tau_{2\sigma})} dt$$

$$p_{2\sigma+2}^{(1/2;1/2)} = \frac{1}{\pi i} \int_{\tau_{2\sigma}\tau_{2\sigma+1}} \sqrt{(t-a)(t-b)} \frac{t-\tau_{2\sigma}}{\tau_{2\sigma+2}-\tau_{2\sigma}} dt$$

$$(\sigma = 1,2,\dots.n-1).$$

§2.4 The case of the full integral equation

Let us consider the full first-kind singular integral equation

$$\frac{1}{\pi i} \int_L \frac{\varphi(t)dt}{t-t_0} + \int_L h(t_0,t)\varphi_0(t)dt = f(t_0) \qquad (2.4.1)$$

Below everywhere, let us assume still that we are seeking a solution to equation (2.4.1) that is bounded at the endpoint a and is unbounded at the endpoint b. Then, as mentioned several times, equation (2.4.1) takes the following form:

$$\varphi(t_0) + \frac{1}{\pi i} \int_L \sqrt{\frac{t-a}{t-b}} \frac{\varphi(t)-\varphi(t_0)}{t-t_0} dt +$$

$$+ \frac{1}{\pi i} \int_L \sqrt{\frac{t-a}{t-b}} h(t_0,t)\varphi(t)dt = f(t_0) \qquad (2.4.2)$$

For integral $S^{(1/2;-1/2)}(\varphi,t_0)$ we use approximation $D_n^{(1/2;-1/2)}(\varphi;t_0)$

$$S^{(1/2;-1/2)}(\varphi,t_0) \approx D_n^{(1/2;-1/2)}(\varphi;t_0)$$

constructed in §15. As for regular part, for it we use a quadrature process which is based on piece-wise quadratic interpolation L_n of the integrand given in §1.4 (for the same nodes), thus, suppose that

$$\int_L h(t_0,t)\varphi_0(t)dt \approx \int_L L_n[h(t_0,t)\varphi_0,t)] dt$$

which represents a quadrature sum defined by the known coefficients. Consequently, along with equivalent to (2.4.2) equation we will discuss the following equation

$$\varphi(t_0) + \left[S^{(1/2;-1/2)}\right]^{-1} \int_L h(t_0, t)\varphi(t)dt = \left[S^{(1/2;-1/2)}\right]^{-1} f,$$

in space C.

Let us change the left hand side of the initial equation

$$D_n^{(1/2;-1/2)}(\varphi; t_0) + L_n\left[\int_L (L_n(h(t_0, t)\varphi)(t)dt; t_0\right].$$

Take into account that under the above assumptions, $D_n^{(1/2;-1/2)}$ is invertible. Then, if we utilize the piecewise interpolation approximation for the function f, we can write:

$$\varphi(t_0) + \left[D_n^{(1/2;-1/2)}\right]^{-1}\left[L_n\left[\int_L (L_n h(t_0, t)\varphi_n, t_0\right]dt\right].$$
$$= \left[D_n^{(1/2;-1/2)}\right](L_n f)(t_0).$$

If we take into account the structure of the operator $D_n^{(1/2;-1/2)}$, it becomes evident that the operator on the left-hand side of the last equation acts from the piecewise polynomial subspace (of degree ≤ 2) into the same subspace in which the equation is being considered.

Let us show that the following equation

$$D_n^{(1/2;-1/2)}(\varphi; t_0) + L_n\left[\int_L (L_n h(t_0, t)\varphi_n(t)dt; t_0\right]. = L_n[f; t_0] \qquad (2.4.2')$$

has a unique solution for sufficiently big n.

Denote

$$(H\varphi)(t_0) = \left(\int_L h(t_0, t)\varphi(t)dt\right)(t_0)$$

$$(H_n\varphi)(t_0) = \left(L_n \int_L (L_n h(t_0, t)\varphi)(t)dt\right)(t_0),$$

We begin with consideration of the following difference

$$\left(S^{(1/2;-1/2)}\right)^{-1}(H\varphi_n) - \left(D_n^{(1/2;-1/2)}\right)^{-1}(H_n\varphi_n) =$$

$$= \left(D_n^{(1/2;-1/2)}\right)^{-1}(H\varphi_n - H_n\varphi_n) + \left(S^{(1/2;-1/2)}\right)^{-1}(H\varphi_n) -$$

$$- \left(D_n^{(1/2;-1/2)}\right)^{-1}(H\varphi_n) = \left(D_n^{(1/2;-1/2)}\right)^{-1}(H - H_n)\varphi_n +$$

$$+ \left(D_n^{(1/2;-1/2)}\right)^{-1}\left(D_n^{(1/2;-1/2)} - S^{(1/2;-1/2)}\right)\left(S^{(1/2;-1/2)}\right)^{-1}(H\varphi_n).$$

Noting that $\|L\| = 0(1)$ and using the obtained relation $\left\|\left(D_n^{(1/2;-1/2)}\right)^{-1}\right\| = 0(\ln^2 n) \quad (n \to \infty) \cdot h(t_0, t)$ in the above assumptions, we get

$$\left\|\left(D_n^{(1/2;-1/2)}\right)^{-1}(H\varphi_n - H_n\varphi_n)\right\|_c \leq C_0(n^{-\alpha+\frac{1}{2}} \ln^2 n \|\varphi_n\|$$

$$(C_0 = const) \qquad (n \to \infty).$$

Denote the operator-function $\left[D_n^{(1/2;-1/2)}\right]^{-1}(H\varphi_n)(t_0)$ by $\Psi_n(t_0)$.

We see that

$$\left(S^{(1/2;-1/2)} - D_n^{(1/2;-1/2)}\right)\left[S^{(1/2;-1/2)}\right]^{-1}(H\varphi_n)$$

$$= \left(S^{(1/2;-1/2)} - D_n^{(1/2;-1/2)}\right)\left[S^{(1/2;-1/2)}\right](H\varphi_n)$$

represents the residual term of the quadrature process constructed by us in §1.5

$$S^{(1/2;-1/2)}(\psi_n; t_0) = \frac{1}{\pi i}\int_{ab}\sqrt{\frac{t-a}{t-b}}\frac{\psi_n(t)}{t-t_0}$$

At this, according to the structure of constructing the function $\psi_n(t)$, via the well known reasoning [77], it is clear that $\psi_n(t) \in H_{\alpha-1/2} \quad (1/2 < \alpha \leq 1)$.

Now let us estimate the corresponding residual term

$$R_n(\psi_n; t_0) = \left(S^{(1/2;-1/2)}(\psi_n; t_0) - D_n^{(1/2;-1/2)}(\psi_n; t_0)\right)$$

Let us use the condition $(\psi_n; t) \in H_{\alpha-1/2}$.

Similarly to the previous reasoning, this estimation mainly is reduced to estimation of the residual term $R_n(\varphi; t_0)$ at nodes $\tau_\nu (\nu = 1, 2, \ldots, n)$:

$$R_n(\psi_n; \tau_\nu) = \frac{1}{\pi i} \int_{ab} \sqrt{\frac{t-a}{t-b}} \frac{\psi_n(t) - \varphi_n(\tau_\nu)}{t - \tau_\nu} - D_n^{(1/2;-1/2)}(\varphi_n; \tau_\nu),$$

The latter, having in mind the structure $D_n^{(1/2;-1/2)}(\varphi_n; \tau_\nu)$, can be presented as follows

$$R_n(\psi_n; \tau_\nu) = \sum_{\sigma=1}^{n} \{\frac{1}{\pi i} \int_{\tau_{2\sigma-1}\tau_{2\sigma+1}} \sqrt{\frac{t-a}{t-b}} \frac{\psi_n(t) - \psi_n(\tau_\nu)}{t - \tau_\nu} dt -$$
$$- \left[P_{2\sigma-1}^{(1/2;-1/2)} \frac{\psi_n(\tau_{2\sigma-1}) - \psi_n(\tau_\nu)}{\tau_{2\sigma-1} - \tau_\nu} + P_{2\sigma+1}^{(1/2;-1/2)} \frac{\psi_n(\tau_{2\sigma+1}) - \varphi_n(\tau_\nu)}{\tau_{2\sigma+1} - \tau_\nu} \right] \}$$

$$(2.4.3)$$

Each summand within the curly brackets, which we denote subsequently as $R_n(\psi_n; \tau_\nu)$, represents the residual on the arcs $\tau_{2\sigma-1} \quad \tau_{2\sigma+1}$ for the corresponding integrals derived from the difference relations $\frac{\psi_n(t) - \psi_n(\tau_\nu)}{t - \tau_\nu}$. These residuals are obtained by linear interpolation of the given function using the endpoints of the respective arcs.

As already noted in §1.5, such approximation provides exact values for the function $\psi_n(t)$ that is a polynomial of degree less than or equal to 2. Consequently, the corresponding terms remain unchanged if the divided differences they contain are replaced by any constant. Specifically, any term $\sigma(1 \leq \sigma \leq n)$ in the sum under consideration can be expressed, for instance, as follows:

$$\frac{1}{\pi i} \int_{\tau_{2\sigma-1}\tau_{2\sigma+1}} \sqrt{\frac{t-a}{t-b}} \left[\frac{\psi_n(t) - \psi_n(\tau_\nu)}{t - \tau_\nu} - \frac{\psi_n(\tau_{2\sigma-1}) - \psi_n(\tau_\nu)}{\tau_{2\sigma-1} - \tau_\nu} \right] dt -$$
$$- P_{2\sigma+1}^{(1/2;-1/2)} \left[\frac{\psi_n(\tau_{2\sigma+1}) - \psi_n(\tau_\nu)}{\tau_{2\sigma+1} - \tau_\nu} - \frac{\psi_n(\tau_{2\sigma-1}) - \varphi_n(\tau_\nu)}{\tau_{2\sigma-1} - \tau_\nu} \right].$$

In $R_{n\sigma}$, we subtract a constant of type $\frac{[\psi_n(\tau_{2\sigma+1}) - \psi_n(\tau_\nu)]}{\tau_{2\sigma+1} - \tau_\nu}$ from each devided difference. We represent the latter as:

$$\frac{1}{\pi i}\int_{\tau_{2\sigma-1}\tau_{2\sigma+1}}\left[\frac{\psi_n(t)-\psi_n(\tau_{2\sigma-1})}{t-\tau_v}+(\tau_{2\sigma-1}-t)\frac{\psi_n(\tau_{2\sigma-1})-\psi_n(\tau_v)}{(t-\tau_v)(\tau_{2\sigma-1}-\tau_v)}\right]dt\ -$$

$$-P_{2\sigma+1}^{(1/2;-1/2)}\left[\begin{array}{c}\dfrac{\psi_n(\tau_{2\sigma+1})-\psi_n(\tau_v)}{\tau_{2\sigma+1}-\tau_v}+(\tau_{2\sigma-1}-\tau_{2\sigma+1})\\[2mm]\dfrac{\psi_n(\tau_{2\sigma-1})-\psi_n(\tau_v)}{(\tau_{2\sigma-1}-\tau_v)(\tau_{2\sigma+1}-\tau_v)}\end{array}\right]$$

$$(2.4.4)$$

Let us use the condition $\psi_n \in H_{\alpha-1/2}$, smoothness of contour $L \equiv ab$ and denote $h = \frac{e}{n}$ where l is the length of the line, then we can get convinced that the integral from the given expression does not exceed $O\left(h^{\alpha-1/2}\right)\|\psi_n\|$. Analogous estimations for integral terms of the same expression will take place.

Now let us estimate the sum $\sum_{\sigma=1}^{n} R_{n\sigma}(\psi_n; \tau_v)$, at this under $R_{n\sigma}(\psi_n; \tau_v)$ we mean an expression corresponding to (2.4.4). Similarly to the previous reasonings, we come to the following

$$\sum_{\sigma=1}^{n}\int_{\tau_{2\sigma-1}\tau_{2\sigma+1}}\sqrt{\frac{t-a}{t-b}}\frac{\psi_n(t)-\psi_n(\tau_{2\sigma-1})}{(t-\tau_v)}dt = O\left(h^{\alpha-\frac{1}{2}}\right)\|\psi_n\|_{H_{\alpha-1/2}}$$

$$\sum_{\sigma=1}^{n+1}\frac{1}{\sigma} = O\left(h^{\alpha-1/2}\ln n\right)\|\psi_n\|_{H_{\alpha-1/2}} \quad (n \to \infty).$$

As for sum of integrals

$$(\tau_{2\sigma-1}-t)\frac{\psi_n(\tau_{2\sigma}-1)-\psi_n(\tau_v)}{(\tau_{2\sigma-1}-\tau_v)(t-\tau_v)}$$

from representations of type

$$(\tau_{2\sigma-1}-t)\frac{\psi_n(\tau_{2\sigma}-1)-\psi_n(\tau_v)}{(\tau_{2\sigma-1}-\tau_v)(t-\tau_v)},$$

for them we will have

$$O\left(h^{\alpha-1/2}\right)\|\psi_n\|_{H_{\alpha-1/2}}\sum_{\sigma=1}^{n+1}\frac{1}{\sigma^{2/3-\alpha}} \quad (1/2 < \alpha < 1),$$

from where we get estimation $O\left(h^{\alpha-1/2}\right)\|\psi_n\|_{H_{\alpha-1/2}}$.

Similar estimations can be derived from the sum of the corresponding non-integral terms, at this all estimates are uniform with respect to v ($1 \le v \le n+1$).

On the basis of these estimates we get estimation $\max\limits_{v}|R_n(\psi_n; \tau_v)|$, which in its turn, as in §1.5, gives the following

$$\left\|\left(S^{(1/2;-1/2)} - D_n^{(1/2;-1/2)}\right)\psi_n\right\| \le \varepsilon_n\|\psi_n\|_{H_{\alpha-1/2}}.$$

with totally definite estimation of ε_n.

Here, as previously,

$$\psi_n(t_0) = \left[S^{(1/2;-1/2)}\right]^{-1}(H\varphi_n)(t_0) = S^{(1/2;-1/2)}(H\varphi_n)(t_0)$$

we have

$$\|\psi_n\|_{H_\beta} \le C_0\|H\varphi_n\|_{H_\beta} \le C_1 \max\limits_{j}|\varphi(\tau_j)|, \quad C_0, C_1 = const.$$

Providing this, also from explicit estimation of ε_n and $\left\|\left[D_n^{(1/2;-1/2)}\right]^{-1}\right\| = O\left(n^{\frac{1}{2}}\ln^2 n\right)$, finally we come to

$$\left\|\left[D_n^{(1/2;-1/2)}\right]^{-1}(D_n^{(1/2;-1/2)} - S^{(1/2;-1/2)}\left[S^{(1/2;-1/2)}\right]^{-1}(H\varphi_n)\right\| \le$$

$$\le \delta_n\|\varphi_n\|, \delta_n \to 0 \ (n \to \infty) \tag{2.4.5}$$

Thus, let us use

$$\left\|\left[S^{(1/2;-1/2)}\right]^{-1}(H\varphi_n) - \left[D_n^{(1/2;-1/2)}\right]^{-1}(H_n\varphi_n)\right\|$$

and take into account the fact that $L_n\left[D_n^{(1/2;-1/2)}\right]^{-1} = \left[D_n^{(1/2;-1/2)}\right]^{-1}$ we obtain similar estimation for $L_n\left[S^{(1/2;-1/2)}\right]^{-1}(H\varphi_n) - \left[D_n^{(1/2;-1/2)}\right]^{-1}(H_n\varphi_n)$.

Based on the reasoning provided above and straightforward estimations (considering $f \in H_\alpha$ ($1/2 < \alpha \le 1$)), we can get convinced that the general conditions [38] for approximation method for the considered equation are fulfilled.

In our assumptions, that the corresponding roots of the polynomial obtained via computer calculations, coincide with the points $t_{0j} = \frac{\tau_j + \tau_{j+1}}{2}$, except for one root that slightly deviates from the point t_{0n}, it can be demonstrated that equation (2.4.2') is uniquely solvable starting from a certain n. Furthermore, the approximate solution tends to the exact one.

In the case of the corresponding smoothness of the functions $h(t_0, t)$ and $f(t_0)$, the order of convergence $O\left(n^{-3/2} \ln n\right)$ is obtained.

We can conduct a similar reasoning (with corresponding refinements) when seeking a solution of equation (2.4.1) that is either unbounded at both ends or bounded at both ends.

Chapter III. Approximate calculation of Cauchy-type integrals and their derivatives for open contours using a corrective parameter

The solution of many problems in mechanics is facilitated using Cauchy-type integrals and their derivatives.

Similarly, in fundamental boundary problems of the theory of elasticity, components of stress and displacement are expressed using Cauchy-type integrals through the formulas of N. Muskhelishvili and V. Kolosov [77].

Therefore, one of the most significant issues is the approximate calculation of such integrals with high accuracy.

§3.1 Construction of a quadrature formula for the approximate calculation of the Cauchy-type integral and its derivative in the case of an open contour

When solving various applied problems numerically (using methods of integral equations), it becomes necessary to calculate approximately following types of integrals

$$F(\rho(t); \varphi; z) \equiv \frac{1}{2\pi i} \int_\Gamma \rho(t) \frac{\varphi(t)}{t - z} \, dt \qquad (3.1.1)$$

and their derivatives [66, 77, 82], where $L \equiv ab$ is some smooth open contour in the complex plane with endpoints a and b, where

$$\rho(t) = (t - a)^\alpha (t - b)^\beta, 0 < |\alpha|, |\beta| < 1$$

is a weight function, under which some fixed branch of this multivalued function is meant. In applied problems, situations often arise when $\alpha, \beta = \pm 1/2$ [66,77,82]. Below, we everywhere assume that $\alpha = -\beta = 1/2$ (the cases for $\alpha = \beta = 1/2$ and $\alpha = \beta = -1/2$ are considered similarly).

Such integrals, for any fixed $z \notin L$ point, can approximately be calculated using conventional quadrature formulas. However, their accuracy decreases as z

no matter how close approaches the points on the contour of integration L (especially its endpoints).

In [102], for Cauchy type integrals and their derivatives for the case of closed smooth contours, approximate quadrature formulas are derived and corresponding estimates for their accuracy are obtained.

Construction of similar formulas for Cauchy-type weighted integrals for the case of an open smooth contour is presented below.

Let us consider a smooth open curve L in the complex plane, whose parameterization is given as $t = t(s)$ $(a \le s \le b)$, We assume that positive direction is taken with respect to the growth of the parameter s.

For any natural n, divide the segment $[a, b]$ into n equal parts using the points:

$$S_\sigma = \frac{b - a}{n}(\sigma - 1) + a \qquad (\sigma = 1,2,\dots,n + 1)$$

and denote $\tau_\sigma = t(s_\sigma)$. Suppose $L_v(\varphi; t_0) = l_{v0}(t_0)\varphi(\tau_v) + l_{v1}(t_0)\,\varphi(\tau_{v+1})$ is a Lagrange interpolation polynomial and consequently

$$l_{v0}(t) = \frac{t - \tau_{v+1}}{\tau_v - \tau_{v+1}}, \quad l_{v1}(t) = \frac{t - \tau_v}{\tau_{v+1} - \tau_v}.$$

Let's assume that t_0 is an arbitrary point $t_0 \in L$ (which will later be associated with a point z in a certain sense), v $(1 \le v \le n)$ is a number for which $t_0 \in \tau_v\tau_{v+1}$. For each t, from the arc $\tau_\sigma\tau_{\sigma+1}$ $(1 \le \sigma \le n)$, we will approach the function $\varphi(t)$ in the following manner:

when $\sigma = v - 1$

$$\varphi(t) \approx L_v(\varphi; t_0)$$
$$+ (t - t_0)\left\{ l_{v-10}(t)\frac{\varphi(\tau_{v-1}) - \varphi(t_0)}{\tau_{v-1} - t_0}\right.$$
$$\left. + l_{v-11}(t)\frac{\varphi(\tau_v) - L_v(\varphi; t_0)}{\tau_v - t_0}\right\};$$

for $\sigma = v$

$$\varphi(t) \approx L_v(\varphi; t_0)$$
$$+ (t - t_0)\left\{l_{v0}\frac{\varphi(\tau_v) - L_v(\varphi; t_0)}{\tau_v - t_0} + l_{v1}(t)\frac{\varphi(\tau_{v+1}) - L_v(\varphi; t_0)}{\tau_{v+1} - t_0}\right\};$$

for $\sigma = v + 1$
$$\varphi(t) \approx L_v(\varphi; t_0)$$
$$+ (t - t_0)\left\{l_{v+10}\frac{\varphi(\tau_{v+1}) - L_v(\varphi; t_0)}{\tau_{v+1} - t_0}\right.$$
$$\left. + l_{v+11}(t)\frac{\varphi(\tau_{v+2}) - \varphi(t_0)}{\tau_{v+2} - t_0}\right\};$$

We introduce the notation

$$L_{\sigma k}(\varphi; t_0) = \begin{cases} \varphi(t_0), & \sigma + k \neq v, v + 1; \\ L_v(\varphi; t_0), & \sigma + k = v, v + 1, (t_0 \in \tau_v \tau_{v+1}) \end{cases}$$

Then generally we can write

$$\varphi(t) \approx L_v(\varphi; t_0) + (t - t_0) \sum_{k=0}^{1} l_{\sigma k}(t)\frac{\varphi(\tau_{\sigma+k}) - L_{vk}(\varphi; t_0)}{\tau_{\sigma+k} - t_0} \qquad (3.1.2)$$

Substitute this representation of $\varphi(t)$ in the expression of $F(p; \varphi; z)$ and denote

$$p_{\sigma k}(\rho; t_0; z) = \frac{1}{2\pi i}\int_{\tau_\sigma \tau_{\sigma+1}} \sqrt{\frac{t - a}{t - b}}\frac{(t - t_0)l_{\sigma k}(t)}{t - z}\, dt \quad (\sigma = 1, 2, \dots, n),$$

we will have

$$F(\rho; \varphi; z) = F(\rho; 1; z)L_v(\varphi; t_0)$$
$$+ \sum_{\sigma=1}^{n}\sum_{k=0}^{1} p_{\sigma k}(\rho; t_0; z)\frac{\varphi(\tau_{\sigma+k}) - L_{\sigma k}(\varphi; t_0)}{\tau_{\sigma+k} - t_0} +$$

$$+ \frac{1}{2\pi i}\sum_{\sigma=1}^{n}\int_{\tau_\sigma \tau_{\sigma+1}} \frac{R_{n\sigma}(\rho; \varphi; t; t_0)}{t - z}\, dt \qquad (t_0 \in \tau_v \tau_{v+1}), \qquad (3.1.3)$$

where

$$p_{\sigma k}(\rho; t_0; z) = \frac{1}{2\pi i} p_\sigma(\rho) + \frac{z - t_0}{2\pi i} \int_{\tau_\sigma \tau_{\sigma+1}} \sqrt{\frac{t-a}{t-b}} \frac{l_{\sigma k}(t) dt}{t-z}$$

$$p_\sigma(\rho) = \frac{1}{2\pi i} \int_{\tau_\sigma \tau_{\sigma+1}} \sqrt{\frac{t-a}{t-b}} \, dt \, , \ (\sigma = 1, 2, \dots, n) \, ,$$

while $R_{n\sigma}(\rho; \varphi; t; t_0)$ represents the growth of the approximation of $\varphi(t)$ according to formula (3.1.2).

Now let's consider the issue of approximation of the derivative of the integral $F(\rho; \varphi; z)$:

$$F'(\rho; \varphi; z) = \frac{1}{2\pi i} \int_\Gamma \sqrt{\frac{t-a}{t-b}} \frac{\varphi(t)}{(t-z)^2} dt$$

Approach $\varphi(t)$ according to the following scheme:

$$\varphi(t) \approx \varphi(t_0) + (t - t_0)\varphi(t_0, t_1) + (t - t_0)(t - t_1)$$

$$\sum_{k=0}^{1} l_{\sigma k}(t) \frac{\varphi(\tau_{\sigma+k}, t_1) - \varphi(t_0, t_1)}{\tau_{\sigma+k} - t_0} \tag{3.1.4}$$

Where, until now, $t_0, t_1 \in \Gamma$ $(t_1 \neq t_0)$ are the points arbitrarily selected on the curve Γ, and $\varphi(t_0, t_1), \varphi(\tau_{\sigma+k}, t_1)$ are the first-order divided differences. Based on (3.1.3), we will use an analogous approximation for the approximation of $F'(\rho; \varphi; z)$, namely, when $\sigma + k = v, v + 1$, then we change the divided difference

$$\varphi(\tau_{\sigma+k}, t_1) = \frac{\varphi(\tau_{\sigma+k}) - \varphi(t_1)}{\tau_{\sigma+k} - t_1}$$

by expression $\dfrac{\varphi(\tau_{\sigma+k}) - L_v(\varphi; t_1)}{\tau_{\sigma+k} - t_1} \cdot (\tau_{\sigma+k} \neq t_1)$ and relation $\varphi(t_0, t_1) = \dfrac{L_v(\varphi; t_1) - L_v(\varphi, t_0)}{t_1 - t_0}$ holds after which we get:

$$\varphi(t) \approx \varphi(t_0) + (t - t_0)\frac{L_v(\varphi; t_1) - L_v(\varphi; t_0)}{t_1 - t_0} +$$

$$\left\{ +(t-t_0)(t-t_1)\sum_{\kappa=0}^{1}\frac{l_{\sigma\kappa}(t)}{\tau_{\sigma+\kappa}-t_0}\frac{\varphi(\tau_{\sigma+\kappa})-L_{\sigma\kappa}(\varphi;t_1)}{\tau_{\sigma+\kappa}-t_1}-\frac{L_v(\varphi;t_1)-L_v(\varphi;t_0)}{t_1-t_0}\right\}$$

$$(3.1.5)$$

where it is meant that $t\in\tau_\sigma\tau_{\sigma+1}$, $t_0,t_1\in\tau_v\tau_{v+1}$ $\quad(v,\sigma=\overline{1,n})$ and for now $t_0\neq t_1$., insert (3.1.5) into expression of $F'(\rho,\varphi;z)$, we come to

$$F'(\rho;\varphi;z)\approx F'(\rho;1;z)\varphi(t_0)+F'(\rho;t-t_0;z)\{L_v(\varphi;t_1)-L_v(\varphi;t_0)\}+$$

$$+\sum_{\sigma=1}^{n}\sum_{\kappa=0}^{1}p_{\sigma\kappa}\,(\rho;t_0,t_1,z)\frac{1}{\tau_{\sigma+\kappa}-t_0}$$
$$\left\{\frac{\varphi(\tau_{\sigma+\kappa})-L_{\sigma\kappa}(\varphi;t_1)}{\tau_{\sigma+\kappa}-t_1}-\frac{L_v(\varphi;t_1)-L_v(\varphi;t_0)}{t_1-t_0}\right\},$$

$$(3.1.6)$$

where $\qquad p_{\sigma\kappa}=\dfrac{1}{2\pi i}\displaystyle\int_{\tau_\sigma\tau_{\sigma+1}}\sqrt{\dfrac{t-a}{t-b}}\dfrac{(t-t_1)(t-t_0)}{(t-z)^2}dt.$

Here, also we mean that $t\in\tau_\sigma\tau_{\sigma+1}$; $t_0,t_1\in\tau_v\tau_{v+1}$; $t_0\neq t_1$, at this

$$F'(\rho;1;z)=\frac{1}{2\pi i}\int_\Gamma\sqrt{\frac{t-a}{t-b}}\frac{dt}{(t-z)^2},$$

$$F'(\rho;t-t_0;z)=\frac{1}{2\pi i}\sqrt{\frac{t-a}{t-b}}\frac{(t-t_0)dt}{(t-z)^2}$$

In (3.1.3) and (3.1.6), transform the terms, for which $\sigma+\kappa=v,\ v+1$. From (3.1.3) we find

$$\frac{\varphi(\tau_v)-L_v(\varphi;t_0)}{\tau_v-t_0}=\frac{(t_0-\tau_v)[\varphi(\tau_v)-\varphi(\tau_{v+1})]}{(\tau_{v+1}-\tau_v)(\tau_v-t_0)},$$

consequently

$$\frac{\varphi(\tau_v)-L_v(\varphi;t_0)}{\tau_v-t_0}=\frac{\varphi(\tau_{v+1})-\varphi(\tau_v)}{\tau_{v+1}-\tau_v}.$$

In the case when $\sigma = v+1, \kappa = 0$ and $\sigma = v, \kappa = 1$, that is when $\sigma + \kappa = v + 1$, we have

$$\frac{\varphi(\tau_{v+1}) - L_v(\varphi; t_0)}{\tau_v - t_0} = \frac{\varphi(\tau_{v+1}) - \varphi(\tau_v)}{\tau_{v+1} - \tau_v}.$$

On the basis of the above said, we can represent (3.1.3) in the following form

$$F(\rho; \varphi; z) \approx F(\rho; 1; z) L_v(\varphi; t_0) + p_{v-10}(\rho; t_0; z)\frac{\varphi(\tau_{v-1}) - \varphi(t_0)}{\tau_{v-1} - t_0} +$$

$$+[p_{v-11}(\rho, t_0, z) + p_{v0}(\rho, t, z) + p_{v1}(\rho, t_0, z) + p_{v+10}(\rho, t_0, z)] \times$$

$$\times \frac{\varphi(\tau_{v+1}) - \varphi(\tau_v)}{\tau_{v+1} - \tau_v} +$$

$$+p_{v+1,1}(\rho, t_0, z)\frac{\varphi(\tau_{v+2}) - \varphi(t_0)}{\tau_{v+2} - t_0} + \sum_{\sigma \neq v\pm 1, v}^{n} \sum_{k=1}^{1} p_{\sigma k}(\rho, t_0, z)\frac{\varphi(\tau_{v+k}) - \varphi(t_0)}{\tau_{v+k} - t_0}.$$

As is obvious, the right hand side is defined also explicitly when $t_0 = \tau_v, t_0 = \tau_{v+1}$.

Now let's consider (3.1.6). გვაქვს We have

$$\frac{L_v(\varphi; t_0) - L_v(\varphi; t_1)}{t_0 - t_1} = \frac{1}{t_0 - t_1}\left\{\frac{t_0 - t_1}{\tau_v - \tau_{v+1}}\varphi(\tau_v) + \frac{t_0 - t_1}{\tau_{v+1} - \tau_v}\varphi(\tau_{v+1})\right\}$$
$$= \frac{\varphi(\tau_{v+1}) - \varphi(\tau_v)}{\tau_{v+1} - \tau_v}.$$

Besides, using the above resoning,

$$\frac{\varphi(\tau_v) - L_v(\varphi; t_1)}{\tau_v - t_1} = \frac{\varphi(\tau_{v+1}) - \varphi(\tau_v)}{\tau_{v+1} - \tau_v}.$$

Following from this, for any $t_0, t_1 \in \Gamma$, considered above, we have:

$$\left\{\frac{\varphi(\tau_v) - L_v(\varphi; t_1)}{\tau_v - t_1} - \frac{L_v(\varphi; t_0) - L_v(\varphi; t_1)}{t_0 - t_1}\right\}\frac{1}{\tau_v - t_0} = 0$$

$$\left\{ \frac{\varphi(\tau_{\nu+1}) - L_\nu(\varphi; t_1)}{\tau_{\nu+1} - t_1} - \frac{L_\nu(\varphi; t_1) - L_\nu(\varphi; t_0)}{t_1 - t_0} \right\} \frac{1}{\tau_{\nu+1} - t_0} = 0.$$

Consequently (3.1.6) can be presented as

$$F'(\rho; \varphi; z) \approx F'(\rho; 1; z) L_\nu(\varphi; t_0) + F'(\rho; t - t_0; z) \frac{\varphi(\tau_{\nu+1}) - \varphi(\tau_\nu)}{\tau_{\nu+1} - \tau_\nu} +$$

$$+ \frac{p_{\nu-10}(\rho; t_0; t_1; z)}{\tau_{\nu-1} - t_0} \left\{ \frac{\varphi(\tau_{\nu-1}) - \varphi(t_1)}{\tau_{\nu-1} - t_1} - \frac{\varphi(\tau_{\nu+1}) - \varphi(\tau_\nu)}{\tau_{\nu+1} - \tau_\nu} \right\} +$$

$$+ \frac{p_{\nu+11}(\rho, t_0, t_1, z)}{\tau_{\nu+2} - t_0} \left\{ \frac{\varphi(\tau_{\nu+2}) - \varphi(t_1)}{\tau_{\nu+2} - t_1} - \frac{\varphi(\tau_{\nu+1}) - \varphi(\tau_\nu)}{\tau_{\nu+1} - \tau_\nu} \right\} +$$

$$+ \sum_{\sigma \neq \nu, \nu \pm 1;\ \kappa=0}^{n} \sum_{}^{1} \frac{p_{\sigma\kappa}(\rho, t_0, t_1, z)}{\tau_{\nu+2} - t_0} \times \left\{ \frac{\varphi(\tau_{\nu+\kappa}) - \varphi(t_1)}{\tau_{\nu+\kappa} - t_1} - \frac{\varphi(\tau_{\nu+1}) - \varphi(\tau_\nu)}{\tau_{\nu+1} - \tau_\nu} \right\}$$

$$(3.1.6')$$

In the last expression $t_0, t_1 \in \tau_\nu \tau_{\nu+1}$ (at this we can assume that they coinside with nodes $\tau_\nu, \tau_{\nu+1}$). Further we will assume everywhere that $t_0 = t_1$. Consequintly, the coefficients $p_{\sigma\kappa}(\rho, t_0, t_1, z)$ will have the following form:

$$p_{\sigma k}(\rho, t_0, z) = \frac{1}{2\pi\iota} \int_{\tau_\sigma \tau_{\sigma+1}} \sqrt{\frac{t - a}{t - b}} \frac{(t - t_0)^2 l_{\sigma\kappa}(t) dt}{(t - z)^2} =$$

$$= \frac{1}{2} p_\sigma(\rho) + \frac{1}{2\pi\iota} \int_{\tau_\sigma \tau_{\sigma+1}} \sqrt{\frac{t - a}{t - b}} \frac{(t - t_0)^2 - (t - z)^2}{(t - z)^2} l_{\sigma\kappa}(t) dt =$$

$$= \frac{1}{2} p_\sigma(\rho) + \frac{1}{2\pi\iota} \int_{\tau_\sigma \tau_{\sigma+1}} \sqrt{\frac{t - a}{t - b}} \frac{(z - t_0)(t_0 - 2t + z)}{(t - z)^2} l_{\sigma\kappa}(t) dt = \frac{1}{2} p_\sigma(\rho) +$$

$$+ \frac{(z - t_0)}{2\pi\iota} \left\{ \int_{\tau_\sigma \tau_{\sigma+1}} \sqrt{\frac{t - a}{t - b}} \frac{(t - t_0)}{(t - z)^2} l_{\sigma\kappa}(t) dt + \int_{\tau_\sigma \tau_{\sigma+1}} \sqrt{\frac{t - a}{t - b}} \frac{t - z}{(t - z)^2} l_{\sigma\kappa}(t) dt \right\}$$

$$= \frac{1}{2} p_\sigma(\rho) +$$

$$+\frac{(z-t_0)}{2\pi\iota}\left\{\int_{\tau_\sigma\tau_{\sigma+1}}\sqrt{\frac{t-a}{t-b}}\frac{t-z}{(t-z)^2}l_{\sigma\kappa}(t)dt+(z-t_0)\int_{\tau_\sigma\tau_{\sigma+1}}\sqrt{\frac{t-a}{t-b}}\frac{l_{\sigma\kappa}(t)}{(t-z)^2}dt\right.$$

$$\left.+\int_{\tau_\sigma\tau_{\sigma+1}}\sqrt{\frac{t-a}{t-b}}\frac{l_{\sigma\kappa}(t)dt}{t-z}\right\}$$

$$=\frac{1}{2}p_\sigma(\rho)+\frac{(z-t_0)}{\pi\iota}\int_{\tau_\sigma\tau_{\sigma+1}}\sqrt{\frac{t-a}{t-b}}\frac{l_{\sigma\kappa}(t)dt}{t-z}+$$

$$+\frac{(z-t_0)^2}{2\pi\iota}\int_{\tau_\sigma\tau_{\sigma+1}}\sqrt{\frac{t-a}{t-b}}\frac{l_{\sigma\kappa}(t)dt}{(t-z)^2}.$$

§3.2 Estimation of the Residual Term of the Quadrature Formula which Approximates Cauchy-Type Integrals

Let us now evaluate the approximation order of integrals $F(\rho;\varphi;z)$ and $F'(\rho;\varphi;z)$ when replaced by their respective approximations.

To achieve this, we use the integral representation of the residual term when approximating $\varphi(t)$. The corresponding residual term $R_{n\sigma}(\rho;\varphi;t;t_0;)$ defined by (3.1.3) will have the following form:

$$[\varphi(t_0)-L_v(\varphi;t_0)]+(t-t_0)r_{n\sigma}(\varphi;t;t_0),$$

where

$$r_{n\sigma}(\varphi;t;t_0)=-\sum_{\kappa=0}^{1}l_{\sigma\kappa}(t)\frac{\varphi(\tau_{\sigma+\kappa})-L_{\sigma\kappa}(\varphi;t_0)}{\tau_{\sigma+\kappa}-t_0}.$$

Let us assume that the function $\varphi(t)$ has a second-order bounded derivative $\varphi''(t)$ on the line Γ. We us use the Taylor expansion for the first term enclosed in square brackets,

$$\varphi(t_0)=\varphi(t_v)+(t_0-\tau_v)\varphi'(\tau_v)+$$

$$+\int_{\tau_v t_0}(t_0-u)\varphi''(u)du-l_{v1}(t_0)\int_{\tau_v\tau_{v+1}}(\tau_{v+1}-u)\varphi''(u)du,$$

where

$$\varphi(t_0) - L_v(\varphi; t_0) = \int_{\tau_v t_0} (t_0 - u)\varphi''(u)du -$$

$$-l_{v1}(t_0) \int_{\tau_v \tau_{v+1}} (\tau_{v+1} - u)\varphi''(u)du.$$

Accordingly, we will have

$$\varphi(t_0) - L_v(\varphi; t_0) = O(h^2)\sup_{t\in\Gamma}|\varphi''(t)| = O(h^2)M$$

$$\left(M = \sup_{t\in\Gamma}|\varphi''(t)|\right). \tag{3.2.1}$$

Now taking into account (see [82])

$$\int_{-e}^{e} \sqrt{\frac{e \pm t}{e \mp t}} \frac{dt}{t - z_0} = -\frac{\pi(e \pm z_0)}{\sqrt{z_0^2 - e^2}} \pm \pi$$

we can similarly show, that

$$\left|\frac{1}{2\pi i}\int_a^b \sqrt{\frac{t - a}{t - b}} \frac{[\varphi(t_0) - L_v(\varphi; t_0)]dt}{t - z}\right| = O(h^2)\sqrt{\frac{a - z}{b - z}}.$$

Note that for each $\sigma(1 \leq \sigma \leq n)$ $(\sigma + \kappa \neq v, v + 1)$, the value $r_{n\sigma}(\varphi; t, t_0)$ is zero on the arc $\tau_\sigma \tau_{\sigma+1}$, when $\varphi(t) = \varphi_\sigma(t)$ is a polynomial of degree ≤ 2 with respect to t and independently of σ at the point $t = t_0$ holds the value $\varphi(t_0)$: $\varphi_\sigma(t)|_{t=t_0} = \varphi(t_0), \sigma = \overline{1, n}$.

For any given $t \in \tau_\sigma \tau_{\sigma+1}$, $\sigma \neq v \pm 1$, v following expansion for (t) takes place

$$\varphi(t) = \lambda_{\sigma v}(\varphi; t; t_0) + (t - t_0) \times$$

$$\times \left\{2\int_{\tau_\sigma t} \frac{(t - \tau)d\tau}{(t_0 - \tau)^3}\int_{\tau t_0}(t_0 - u)\varphi''du - \int_{\tau_\sigma t} \frac{t - \tau}{t_0 - \tau}\varphi''(\tau)d\tau\right\},$$

$$t \in \tau_\sigma \tau_{\sigma+1}, \qquad (\sigma \neq v, v \pm 1)$$

$$\lambda_{\sigma v}(\varphi; t, t_0) + (t - t_0) = \varphi(t_0) + \frac{t - t_0}{\tau_\sigma - t_0}[\varphi(\tau_\sigma) - \varphi(t_0)] +$$

$$+ \frac{(t - t_0)(t - \tau_\sigma)}{(t_0 - \tau_\sigma)^2} \int_{\tau t_0} (t_0 - u)\varphi''(u)du$$

$$(3.2.2)$$

The Taylor formula provides the corresponding estimate when $|t - t_0| = O(n^{-1})$, i.e., when σ and v are close to each other. Therefore, the expansion (3.2.2) is used here. The obtained expansion is accepted when the expression enclosed in the curly brackets has been changed by expression

$$\int_{\tau_\sigma t} (t - \tau)\left\{\frac{d}{d\tau}\frac{1}{(t_0 - \tau)^2}\int_{\tau t_0}(t_0 - u)\varphi''(u)du\right\}$$

and afterwards the required number of partial integrations has been performed.

Let us note that for any σ $(1 \leq \sigma \leq n)$, $\lambda_{\sigma v}(\varphi; t; t_0)$ represents a polynomial with respect to t, whose degree is less than or equal to two, and which, when $t = t_0$, $(t_0 \in \tau_v \tau_{v+1};\quad 1 \leq v \leq n)$ holds a value $\varphi(t_0)$. Taking this into account, we replace (3.2.2) instead of $\varphi(t)$ in the expression of $r_{n\sigma}(\varphi; t; t_0)$.

At this, counting

$$\sigma \neq v \pm 1, v, \qquad L_{\sigma \kappa}(\varphi; t_0) = \varphi(t_0), \left(\kappa = \overline{0,1}\right),$$

we get

$$r_{n\sigma}(\varphi, t, t_0) = (t - t_0) \times$$

$$\times \left\{2\int_{\tau_\sigma t}\frac{(t - \tau)d\tau}{(t_0 - \tau)^3}\int_{\tau t_0}(t_0 - u)\varphi''(u)\,du - \int_{\tau_\sigma t}\frac{(t - \tau)}{t_0 - \tau}\varphi''(\tau)d\tau\right\} -$$

$$-l_{\sigma 1}(t)\left\{2\int_{\tau_\sigma \tau_{\sigma+1}}\frac{(\tau_{\sigma+1} - \tau)d\tau}{(t_0 - \tau)^3}\int_{\tau t_0}\begin{aligned}(t_0 - u)\varphi''(u)du -\\ \int_{\tau_\sigma \tau_{\sigma+1}}\frac{(\tau_{\sigma+1} - \tau)}{t_0 - \tau}\varphi''(\tau)dt\end{aligned}\right\}$$

$$(3.2.3)$$

In the case when $\sigma = v \pm 1, v$ in the corresponding representations of $\varphi(t)$, we proceed in the same way as during the derivation of (3.1.6').

We will get:

when $t \in \tau_{v-1}\tau_v$

$$r_{nv-1}(\varphi; t, t_0) = \varphi(t, t_0) - L_{v-10}(t)\frac{\varphi(\tau_{v-1}) - \varphi(t_0)}{\tau_{v-1} - t_0} -$$

$$-l_{v-11}(t)\frac{\varphi(\tau_{v+1}) - \varphi(\tau_v)}{\tau_{v+1} - \tau_v};$$

when $t \in \tau_v\tau_{v+1}$

$$r_{nv}(\varphi; t; t_0) = \varphi(t, t_0) - \frac{\varphi(\tau_{v+1}) - \varphi(\tau_v)}{\tau_{v+1} - \tau_v};$$

when $t \in \tau_{v+1}\tau_{v+2}$

$$r_{nv+1}(\varphi; t; t_0) = \varphi(t, t_0) - l_{v+10}(t)\frac{\varphi(\tau_{v+1}) - \varphi(\tau_v)}{\tau_{v+1} - \tau_v} -$$

$$-l_{v+11}(t)\frac{\varphi(\tau_{v+2}) - \varphi(t_0)}{\tau_{v+2} - t_0}.$$

Let us again use the Taylor formula in the following form:

$$\varphi(t) = \varphi(t_0) + (t - t_0)\varphi'(t_0) + \int_{t_0 t}(\iota - u)\varphi''(u)du.$$

Due to the fact that r_{nj} $(j = v \pm 1, v)$ becomes zero for linear functions, we will have

$$r_{nv-1}(\varphi; t; t_0) = \frac{1}{t - t_0}\int_{t_0 t}(t - u)\varphi''(u)\,du -$$

$$-\frac{l_{v-10}(t)}{\tau_{v-1} - t_0}\int_{t_0\tau_{v-1}}(\tau_{v-1} - u)\varphi''(u)du - \frac{l_{v-11}}{\tau_{v+1} - \tau_v} \times$$

$$\times \left\{\int_{t_0\tau_{v+1}}(\tau_{v+1} - u)\varphi''(u)du - \int_{t_0\tau_v}(\tau_v - u)\varphi''(u)du\right\}. \quad (t \neq t_0)$$

Similarly, this will hold in the remaining cases (that is, when $j = v + 1, v$)

After this, let us estimate the residual term of $F(\rho; \varphi; z) -$, which in turn is reduced to the estimation of sums of the following type

$$\sum_{\sigma=1}^{n} \int_{\tau_\sigma \tau_{\sigma+1}} \sqrt{\frac{t-a}{t-b}} \, \frac{(t-t_0) r_{n\sigma}(\varphi; t; t_0)}{t-z} dt =$$

$$= \sum_{\sigma=1}^{n} \left\{ \int_{\tau_\sigma \tau_{\sigma+1}} \sqrt{\frac{t-a}{t-b}} \, r_{n\sigma}(\varphi; t; t_0) dt + \right.$$

$$\left. +(z-t_0) \int_{\tau_\sigma \tau_{\sigma+1}} \sqrt{\frac{t-a}{t-b}} \, \frac{r_{n\sigma}(\varphi; t; t_0) dt}{t-z} \right\}.$$

$$\sum_{\sigma=1}^{n} \int_{\tau_\sigma \tau_{\sigma+1}} \sqrt{\frac{t-a}{t-b}} r_{n\sigma}(\varphi; t; t_0) dt = \sum_{\sigma \neq v \pm 1, v} + \sum_{\sigma = v \pm 1, v}$$

We can easily obtain the estimates of the sum if we use the representation (3.2.2) and the values of the expression for $r_{n\sigma}(\varphi; t; t_0)$ at the nodes $\sigma = v \pm 1, v$. Namely,

$$\left| \sum_{\sigma \neq v \pm 1, v} \int_{\tau_\sigma \tau_{\sigma+1}} \sqrt{\frac{t-a}{t-b}} r_{n\sigma}(\varphi; t; t_0) dt \right| =$$

$$= O(1) \sum_{\sigma \neq v \pm 1, v} \int_{S_\sigma}^{S_{\sigma+1}} \left| \sqrt{\frac{t-a}{t-b}} r_{n\sigma}(\varphi; t; t_0) ds \right| =$$

$$= O(h^{3/2}) \sum_{j=1}^{n} \frac{1}{j} O(h^{3/2} \ln n) M \ (n > 1).$$

Uniformly for v and t_0. Similarly, $\left| \sum_{\sigma = v \pm 1, v} \right| = O(h^{3/2})$.

As for the estimation of the integral $(z - t_0) \int_{\tau_\sigma \tau_{\sigma+1}} \sqrt{\frac{t-a}{t-b}} \frac{r_{n\sigma}(\varphi; t; t_0) dt}{t-z}$, we will assume that the parameter t_0, which we have been taking arbitrarily from Γ, is chosen so that for the given z, the distance between z and t_0 does not exceed the minimum distance between z and the points of the line Γ. Then, we can write

$$\left| (z - t_0) \int_{\tau_\sigma \tau_{\sigma+1}} \sqrt{\frac{t-a}{t-b}} \, \frac{r_{n\sigma}(\varphi; t; t_0)}{t-z} \, dt \right|$$

$$= O(1) \int_{S_\sigma}^{S_{\sigma+1}} \left| \sqrt{\frac{t-a}{t-b}} \, r_{n\sigma}(\varphi; t; t_0) \right| ds$$

for every σ. Taking into account the latter, for the sum

$$\sum_\sigma (z - t_0) \int_{\tau_\sigma \tau_{\sigma+1}} \sqrt{\frac{t-a}{t-b}} \, \frac{r_{n\sigma}(\varphi; t; t_0)}{t-z} \, dt$$

we come to analogous estimation.

Now let's estimate the residual term of the integral $F'(\rho; \varphi; z)$. Due to (3.1.6) we will have

$$[\varphi(t_0) - L_v(\varphi; t_0)] F'(\rho; 1; z;) = [\varphi(t_0) - L_v(\varphi; t_0)] \int_{ab} \sqrt{\frac{t-a}{t-b}} \, \frac{dt}{(t-z)^2},$$

If we use the above resoning on ab, due to

$$\frac{1}{\pi \iota} \int_{-e}^{+e} \sqrt{\frac{e \pm t}{e \mp t}} \, \frac{dt}{(t-z_0)^2} = \frac{e}{(z_0 \mp e)\sqrt{z_0{}^2 - e^2}}$$

(see [82]) we easily come to the following:

$$|[\varphi(t_0) - L_v(\varphi; t_0)] F'(\rho; 1; z)| = O(h^2) \frac{1}{(z_0 - b)\sqrt{z_0{}^2 - b^2}}.$$

$$|[\varphi(t_0) - L_v(\varphi; t_0)] F'(\rho; 1; z)| = O(h^2) \frac{1}{(z_0 - a)\sqrt{z_0{}^2 - a^2}}.$$

The remainder of the second term in (3.1.6) will be equal to

$$F'(\rho; t - t_0, z) \left[\varphi(t_0, t_1) - \frac{L_v(\varphi; t_0) - L_v(\varphi; t_1)}{t_0 - t_1} \right] =$$

$$= F'(\rho; t - t_0, z) \left\{ \frac{\varphi(t_0) - L_v(\varphi; t_0) - [\varphi(t_1) - L_v(\varphi; t_1)]}{t_0 - t_1} \right\} (t_0 \neq t_1)$$

Let us again use the representations

$$\varphi(t_0) = \varphi(\tau_v) + (t_0 - \tau_v)\varphi'(\tau_v) + \int_{\tau_v t_0}(t_0 - u)\,\varphi''(u)du,$$

$$\varphi(t_1) = \varphi(\tau_v) + (t_1 - \tau_v)\varphi'(\tau_v) + \int_{\tau_v t_1}(t_1 - u)\varphi''(u)du,$$

Then the corresponding expression will take the form of

$$\frac{1}{t_0 - t_1}\{l_{v0}(t_0)\int_{\tau_v t_0}(t_0 - u)\varphi''(u)du +$$

$$+l_{v1}(t_0)\left[\int_{\tau_v t_0}(t_0 - u)\varphi''(u)du - \int_{\tau_v \tau_{\tau+1}}(\tau_{v+1} - u)\varphi''(u)du\right]-$$

$$-l_{v0}(t_1)\int_{\tau_v t_1}(t_1 - u)\varphi''(u)du - \int_{\tau_v t_1}(t_1 - u)\varphi''(u)\,du\}\qquad(3.2.4)$$

$(t_0 \neq t_1)$ (here, as earlier, we use the equality $l_{v0}(t) + l_{v1}(t) \equiv 1$). The expression (3.2.4) does not exceed by modulus $O(h)M$ ($M = \sup|\varphi''(t)|$) (uniformly with respect to t_0, t_1 and v $(t_0, t_1 \in \tau_v \tau_{v+1})$). In our case, the mentioned addend of the remainder term is accompanied by a multiplier

$$F'_{(\rho,t-t_0,z)} = \frac{1}{2\pi\iota}\int_{ab}\sqrt{\frac{t-a}{t-b}}\frac{(t-t_0)}{(t-z)^2}dt =$$

$$= \frac{1}{2\pi\iota}\int_{ab}\sqrt{\frac{t-a}{t-b}}\frac{(t-z)dt}{(t-z)^2} + \frac{1}{2\pi\iota}(z-t_0)\int_{ab}\sqrt{\frac{t-a}{t-b}}\frac{dt}{(t-z)^2}.$$

Considering the above and based on the reasoning outlined earlier, we can conclude that

$$\left|F'(\rho,t-t_0,z)\left\{\frac{\varphi(t_0-)L_v(\varphi;t_0) - [\varphi(t_1) - L_v(\varphi;t_1)]}{t_0 - t_1}\right\}\right| = O(h)\frac{1}{\sqrt{z^2-b^2}}.$$

Now, let us estimate the remainder term of the remaining part of (3.1.6). This change arises from the formula of (3.1.5), which in turn is primarily based on the approximation of $(t - t_0)(t - t_1)\varphi(t, t_0, t_1)$ by the divided difference $\varphi(t, t_0, t_1)$, with respect to $t \in \tau_\sigma \tau_{\sigma+1}$:

$$\varphi(t, t_0, t_1) \approx \sum_{\kappa=0}^{1} l_{\sigma\kappa}(t)\varphi(\tau_{\sigma+\kappa}, t_0, t_1)$$

$$= \sum_{\kappa=0}^{1} \frac{l_{\sigma\kappa}(t)}{\tau_{\sigma+\kappa} - t_0}\left[\varphi(\tau_{\sigma+\kappa}, t_1) - \varphi(t_1, t_0)\right],$$

$$\varphi(t_1, t_0) \approx \sum_{\kappa=0}^{1} \frac{l_{\sigma\kappa}(t)}{\tau_{\sigma+\kappa} - t_0}\left[\frac{\varphi(\tau_{\sigma+\kappa}) - \varphi(t_1)}{\tau_{\sigma+\kappa} - t_1} - \frac{\varphi(t_1) - \varphi(t_0)}{t_1 - t_0}\right] \approx$$

$$\approx \sum_{k=0}^{1} \frac{l_{\sigma\kappa}(t)}{\tau_{\sigma+\kappa} - t_0}\left[\frac{\varphi(\tau_{\sigma+\kappa}) - L_{\sigma\kappa}(\varphi; t_1)}{\tau_{\sigma+\kappa} - t_1} - \frac{L_v(\varphi; t_1) - L_v(\varphi; t_0)}{t_1 - t_0}\right]$$

$$\left(t \in \tau_\sigma \tau_{\sigma+1}; t_1 \neq t_0\right). \tag{3.2.5}$$

Let us express the remainder term of the approximation (3.2.5) in integral form

$$\varphi(t, t_0) = \lambda^*{}_\sigma(\varphi; t; t_0) + 2 \int_{\tau_\sigma t} \frac{(t - \tau)d\tau}{(t_0 - \tau)^3} \int_{\tau t_0} (t_0 - u)\varphi''(u)du -$$

$$- \int_{\tau_\sigma t} \frac{t - \tau}{t_0 - \tau}\varphi''(\tau)d\tau,$$

where $\sigma \neq v \pm 1, v$, $\varphi(t, t_0) = \frac{\varphi(t) - \varphi(t_0)}{t - t_0}$, and $\lambda_\sigma(\varphi; t; t_0) = \frac{\varphi(\tau_\sigma) - \varphi(t_0)}{\tau_\sigma - t_0} + \frac{t - \tau_\sigma}{(t_0 - \tau_\sigma)^3} \int_{\tau_\sigma t_0} (t_0 - u)\varphi''(u)du.$

Let us change the parameter t_0 by t_1 and examine the difference

$$\varphi(t, t_1) - \varphi(t, t_0) = \lambda_\sigma^*(\varphi; t; t_1) - \lambda_\sigma^*(\varphi; t; t_0) +$$

$$+2\left[\int_{\tau_\sigma t} \frac{(t - \tau)d\tau}{(t_1 - \tau)^3} \int_{\tau t_1} (t_1 - u)\varphi''(u)du - \int_{\tau_\sigma t} \frac{(t - \tau)d\tau}{(t_0 - \tau)^3} \int_{\tau t_0} (t_0 - u)\varphi''(u)du\right] -$$

$$- \int_{\tau_\sigma t} \frac{t - \tau}{t_1 - \tau}\varphi''(\tau)d\tau + \int_{\tau_\sigma t} \frac{t - \tau}{t_0 - \tau}\varphi''(\tau)d\tau =$$

$$= \lambda_\sigma^*(\varphi; t; t_1) - \lambda_\sigma^*(\varphi; t; t_0) +$$

$$+2\int_{\tau_\sigma t}(t-\tau)\left\{\int_{\tau t_0}\left[\frac{t_1-u}{(t_1-\tau)^3}-\frac{t_0-u}{(t_0-\tau)^3}\right]\varphi''(u)\,du +\right.$$

$$\left.+\frac{1}{(t_1-\tau)^3}\int_{t_0 t_1}(t_1-u)\varphi''(u)\,du\right\}d\tau - (t_1-t_0)\int_{\tau_\sigma t}\frac{(t-\tau)\varphi''(u)}{(t_0-\tau)(t_1-\tau)}\,du =$$

$$= \lambda_\sigma^{**}(\varphi; t, t_0; t_1) + \rho_\sigma(\varphi; t; t_0),$$

$$(3.2.6)$$

where
$$\rho_\sigma(\varphi; t; t_0) = 2(t_1-t_0)\int_{\tau_\sigma t}(t-\tau)\times$$

$$\times\left\{\int_{\tau t_0}\left[\frac{1}{(t_1-\tau)^3}-(t_0-u)\frac{(t_0-\tau)^2+(t_1-\tau)^2}{(t_0-\tau)^3(t_1-\tau)^3}\right]\varphi''\,du -\right.$$

$$\left.-\frac{1}{(t_1-t_0)(t_1-\tau)^3}\int_{t_0 t_1}(t_1-u)\varphi''(u)\,du\right\}d\tau -$$

$$-(t_1-t_0)\int_{\tau_\sigma t}\frac{(t-\tau)\varphi''(\tau)\,d\tau}{(t_0-\tau)(t_1-\tau)},$$

$$\lambda_\sigma^{**}(\varphi; t; t_0; t_1) = \lambda_\sigma^{**}(\varphi; t; t_0)$$

Let us return to expression (3.2.5) and represent $(t_1 \neq t_0)$

$$\frac{L_v(\varphi; t_1) - L_v(\varphi; t_0)}{t_1 - t_0} = \frac{\varphi(t_1) - \varphi(t_0)}{t_1 - t_0} - \tilde{\rho}_v(\varphi),$$

For $\tilde{\rho}_v(\varphi)$

$$\tilde{\rho}_v(\varphi) = \varphi(t_0, t_1) - \frac{L_v(\varphi; t_1) - L_v(\varphi; t_0)}{t_1 - t_0},$$

According to (3.2.10), there is an estimate $|\tilde{\rho}_v(\varphi)| = O(h)M$. Based on the above, in (3.2.5) we can write:

$$\sum_{\kappa=0}^{1}\frac{l_{\sigma\kappa}(t)}{\tau_{\sigma+\kappa}-t_0}\left[\frac{\varphi(\tau_{\sigma+\kappa})-L_{\sigma\kappa}(\varphi; t_1)}{\tau_{\sigma+\kappa}-t_1} - \frac{\varphi(t_1)-\varphi(t_0)}{t_1-t_0}+\tilde{\rho}_v(\varphi)\right].$$

In the latter expression, when

$$\sigma + \kappa \neq v, v + 1 \Rightarrow L_{\sigma\kappa}(\varphi; t_j) = \varphi(t_j) \qquad (j = 0,1)$$

we can write

$$\sum_{\kappa=0}^{1} \frac{l_{\sigma\kappa}(t)}{\tau_{\sigma+\kappa} - t_0} \left[\varphi(\tau_{\sigma+\kappa}, t_1) - \varphi(t_1, t_0) + \tilde{p}_v(\varphi) \right],$$

where $\varphi(\tau_{\sigma+\kappa}, t_1), \varphi(t_1, t_0)$ represent the first and second-order divided differences, respectively. The last sum can be replaced with the following sum:

$$\sum_{\kappa=0}^{l} \frac{l_{\sigma\kappa}(t)}{t_1 - t_0} \left[\varphi(\tau_{\sigma+\kappa}, t_1) - \varphi(\tau_{\sigma+\kappa}, t_0) + \frac{t_1 - t_0}{\tau_{\sigma+\kappa} - t_0} \tilde{p}_v(\varphi) \right].$$

We can be convinced of their equivalence by direct calculation. On the other hand, the latter expression is equivalent to the primary sum in (3.2.5), with which the second-order divided difference $\varphi(t, t_0, t_1)$ is approximated.

Our main objective is to estimate the difference

$$r_\sigma(\varphi; t; t_0, t_1) = \varphi(t, t_0, t_1) - \sum_{\kappa=0}^{1} l_{\sigma+\kappa}(t)\varphi(\tau_{\sigma+\kappa}, t_0, t_1)$$

for any given $\sigma = 1, 2, \cdots, n$. Namely, from the above mentioned, when $\sigma \neq v \pm 1, v$, this difference has the following form:

$$\frac{\varphi(t, t_1) - \varphi(t, t_0)}{t_1 - t_0} - \sum_{\kappa=0}^{1} l_{\sigma+\kappa}(t) \left[\frac{\varphi(\tau_{\sigma+\kappa}, t_1) - \varphi(\tau_{\sigma+\kappa}, t_0)}{t_1 - t_0} + \frac{\tilde{p}_v(\varphi)}{\tau_{\sigma+\kappa} - t_0} \right].$$

$$(3.2.7)$$

(here we mean that $t \in \tau_\sigma \tau_{\sigma+1}, t_0, t_1 \in \tau_v \tau_{v+1}, t_1 \neq t_0$). The representation of the corresponding difference will also be required in the cases when $\sigma = v \pm 1, v$, for this purpose, let us use the transformations that were applied during the derivation of (3.1.6). For instance, when $\sigma = v - 1$, we will have $r_{v-1} = r_{v-10} + r_{v-11,}$, where

$$r_{v-10}(\varphi; t; t_0, t_1) = \frac{\varphi(t, t_1) - \varphi(t_0, t_1)}{t - t_0} - $$

$$-\frac{l_{v-10}(t)}{\tau_{v-1}-t_0}\left\{\frac{\varphi(\tau_{v-1})-\varphi(t_1)}{\tau_{v-1}-t_1}-\frac{\varphi(\tau_{v+1})-\varphi(\tau_v)}{\tau_{v+1}-\tau_v}\right\}.$$

Here, the relation

$$\frac{\varphi(t,t_1)-\varphi(t,t_0)}{t_1-t_0}=\frac{\varphi(t,t_1)-\varphi(t_1,t_0)}{t-t_0}.$$

was used.

A similar approach applies when t is replaced by $\tau_\sigma,\tau_{\sigma+\kappa}$. Expanding $\varphi(t,t_1)$ into Taylor series around the point t_1, we get:

$$\varphi(t,t_1)=\varphi'(t_1)+\frac{1}{t-t_1}\int_{t_1t}(t-u)\varphi''(u)du.\quad (t\neq t_1)\qquad (3.2.8)$$

Similarly

$$\varphi(t_0,t_1)=\varphi'(t_1)+\frac{1}{t_0-t_1}\int_{t_1t_0}(t_0-u)\varphi''(u)\,du,$$

from where we get

$$\frac{\varphi(t,t_1)-\varphi(t_0,t_1)}{t-t_0}=$$

$$=\frac{1}{t-t_0}\left\{\frac{1}{t_2-t_1}\int_{t_1t}(t-u)\varphi''(u)du-\frac{1}{t_0-t_1}\int_{t_1t_0}(t_0-u)\varphi''(u)du\right\}$$

$$(t\neq t_0,t_1),$$

$$\varphi(\tau_{v-1},t_1)=\varphi(t_1)+\frac{1}{\tau_{v-1}-t_1}\int_{t_1\tau_{v-1}}(\tau_{v-1}-u)\varphi''(u)du;$$

Further, use the representations

$$\varphi(\tau_v)=\varphi(t_1)+(\tau_v-t_1)\varphi'(t_1)+\int_{t_1\tau_v}(\tau_v-u)\,\varphi''(u)du,$$

$$\varphi(\tau_{v+1})=\varphi(t_1)+(\tau_{v+1}-t_1)\varphi'(t_1)\int_{t_1\tau_{v+1}}(\tau_{v+1}-u)\,\varphi''(u)du$$

we get

$$\varphi(\tau_{v+1}, \tau_v) = \frac{\varphi(\tau_{v+1}) - \varphi(\tau_v)}{\tau_{v+1} - \tau_v} =$$

$$= \varphi'(t_1) + \frac{1}{\tau_{v+1} - \tau_v}\left[\int_{t_1\tau_{v+1}}(\tau_{v+1} - u)\varphi''(u)du - \int_{t_1\tau_v}(\tau_v - u)\varphi''(u)du\right],$$

from where

$$\varphi(\tau_{v-1}, t_1) - \varphi(\tau_{v+1}, \tau_v) = \frac{\varphi(\tau_{v-1}) - \varphi(t_1)}{\tau_{v-1} - t_1} - \frac{\varphi(\tau_{v+1}) - \varphi(\tau_v)}{\tau_{v+1} - \tau_v} =$$

$$= \frac{1}{\tau_{v-1} - t_1}\int_{t_1\tau_{v-1}}(\tau_{v-1} - u)\,\varphi''(u)du -$$

$$- \frac{1}{\tau_{v+1} - \tau_v}\left[\int_{t_1\tau_{v+1}}(\tau_{v+1} - u)\varphi''(u)du - \int_{t_1\tau_v}(\tau_v - u)\varphi''(u)du\right].$$

Thus, for r_{v-10} we have:

$$r_{v-10} = \frac{1}{t - t_0}\left\{\frac{1}{t - t_1}\int_{t_1t}(t - u)\,\varphi''(u)du - \frac{1}{t_0 - t_1}\int_{t_1t_0}(t_0 - u)\varphi''(u)\,du\right\}$$

$$- \frac{l_{v-10}(t)}{\tau_{v-1} - t_0}\left\{\frac{1}{\tau_{v-1} - t_1}\int_{t_1\tau_{v-1}}(\tau_{v+1} - u)\,\varphi''(u)du -\right.$$

$$- \frac{1}{\tau_{v+1} - \tau_v}\left[\int_{t_1\tau_{v+1}}(\tau_{v+1} - u)\,\varphi''(u)du - \int_{t_1\tau_v}(\tau_v - u)\varphi''(u)du\right]\right\}.$$

Introduce a notation

$$\aleph_1(\varphi; t) = \frac{1}{t - t_1}\int_{t_1t}(t - u)\varphi''(u)du.$$

For any fixed $t_1 \in \tau_\sigma \tau_{\sigma+1}$ we have

$$\aleph_1(\varphi; t) - \aleph_1(\varphi; t_0) = \int_{t_0 t} \aleph_1'(\varphi; \tau) d\tau \Rightarrow |\aleph_1(\varphi; t) - \aleph_1(\varphi; t_0)|$$

$$\leq C_0 |t - t_0| \sup_{t \in \tau_\sigma \tau_{\sigma+1}} |\aleph_1'(\varphi; t)|,$$

where for any $t \neq t_1$

$$\aleph_1'(\varphi; t) = \frac{1}{(t - t_1)^2} \int_{t_1 t} (t - u)\varphi''(u) du + \frac{1}{t - t_1} \int_{t_1 t} \varphi''(u)\, du;$$

From this it follows that $|\aleph_1'(\varphi; t)| = O(1)M, \left[M = \sup_{u \in \Gamma} |\varphi''(u)| \right].$

The constant involved in $O(1)$ does not dependend on t_1. Accordingly, the expression $\left| \frac{\aleph_1(\varphi;t)-\aleph_1(\varphi;t_0)}{t-t_0} \right|$ also has an estimate $O(1)M$, in which the constants are independent of t_0 and t_1.

As for expression

$$\frac{l_{v-10}(t)}{\tau_{v-1} - t_0} \left\{ \frac{\varphi(\tau_{v-1}) - \varphi(t_1)}{\tau_{v-1} - t_1} - \frac{\varphi(\tau_{v+1}) - \varphi(\tau_v)}{\tau_{v+1} - \tau_v} \right\}$$

similar estimation can be directly obtained for it, taking into account that $|\tau_{v-1} - t_0| > c_1 n^{-1}, |\tau_{v-1} - t_1| > c_2 n^{-1}$,, where C_1, C_2 are positive constants.

Finally, we will have that $r_{v-10} = O(1)M$ uniformly with respect to t_0, t_1 (z_{v-10} and consequently, $\sigma = v - 1, \kappa = 1$, i.e., when $\sigma + \kappa = v$ then $r_{v-11}(\varphi; t; t_0, t_1) = \varphi(t, t_0, t_1) = 0$, for which, as we demonstrated above, the same estimate remains valid. Similarly, the estimates remain valid when $\sigma = v \pm 1$). These estimates hold when $\sigma = v, v + 1$. As for the expression for $r_{v-11}(\varphi; t, t_0, t_1)$, it is exactly equal to $\varphi(t, t_0, t_1)$.

Now, let us derive the integral representation for $r_\sigma(\varphi; t; t_0, t_1) \quad \sigma \neq v \pm 1, v$. Denote $r_\sigma = r_\sigma^* + \frac{\tilde{p}_v(\varphi)}{\tau_{\sigma+\kappa} - t_0}$, the corresponding representation should also be applied to z_σ^*.

In (3.2.7), let us use the expansion from (3.2.6) on both sides and take into account the fact that, regardless of the function $\varphi(t)$, the function

$\frac{\lambda_\sigma^{**}(\varphi;t;t_0,t_1)}{t_0-t_1}$ $(t_0 \neq t_1)$ on the corresponding arcs is linear function with respect to parameter t.

As a result (because the difference (3.2.7) is zero for such functions), from (3.2.6) and (3.2.7) we obtain:

$$r_\sigma{}^*(\varphi;t,t_0,t_1) = \rho_\sigma^*(\varphi;t,t_0,t_1) - \sum_{\kappa=0}^{1} l_{\sigma\kappa}(t)\rho_\sigma^*\big(\varphi;\tau_{\sigma+\kappa},t_0,t_1\big),$$

$$(\sigma \neq v \pm 1, v)$$

where

$$\rho_\sigma^*(\varphi;t;t_0,t_1) = \frac{1}{t_1-t_0}\rho_\sigma(\varphi;t;t_0,t_1) =$$

$$= 2\int_{\bar\tau_\sigma t}(t-\tau)\Big\{\int_{\tau t_0}\Big[\frac{t_1-u}{(t_1-\tau)^3} - \frac{t_0-u}{(t_0-\tau)^3}\Big]\varphi''(u)du - \frac{1}{(t_1-\tau)^3(t_1-t_0)} \times$$

$$\times \int_{t_0 t_1}(t_1-u)\varphi''(u)du\Big\}d\tau - \int_{\tau_\sigma t}\frac{(t-\tau)\varphi''(\tau)d\tau}{(t_0-\tau)(t_1-\tau)} \quad (t_0 \neq t_1).$$

Then, these and the above-obtained representations will be used for the estimation of the remainder term $R_n(\rho;\varphi;z)$

$$R_n(\rho;\varphi;z) = F(\rho;1;z)R^*(\rho,z) + \frac{1}{2\pi i}\sum_{\sigma=1}^{n}\int_{\tau_\sigma\tau_{\sigma+1}}\sqrt{\frac{t-a}{t-b}} \times$$

$$\times (t-t_0)(t-t_1)\frac{r_\sigma(\varphi;t;t_0,t_1)}{(t-z)^2}dt.$$

Represent the integrals on the arcs $\tau_\sigma\tau_{\sigma+1}$ in the form

$$\int_{\tau_\sigma\tau_{\sigma+1}}\sqrt{\frac{t-a}{t-b}}\,(t-t_0)(t-t_1)\frac{r_\sigma(\varphi;t;t_0,t_1)}{(t-z)^2} =$$

$$= \int_{\tau_\sigma \tau_{\sigma+1}} \sqrt{\frac{t-a}{t-b}} \; r_\sigma(\varphi; t, t_0, t_1) dt$$

$$+ \int_{\tau_\sigma \tau_{\sigma+1}} \sqrt{\frac{t-a}{t-b}} \frac{(t-t_0)(t-t_1) - (t-z)^2}{(t-z)^2} \times$$

$$\times r_\sigma(\varphi; t, t_0, t_1) dt,$$

We have

$$(t-t_0)(t-t_1) - (t-z)^2 = (t-t_0)[(t-t_1) - (t-z)] + (t-z)$$

$$[(t-t_0) - (t-z)] = (t-t_0)(z-t_1) + (z-t_0)(t-z)$$

and consequently

$$\frac{(t-t_0)(t-t_1) - (t-z)^2}{(t-z)^2} = \frac{(t-t_0)(z-t_1)}{(t-z)^2} + \frac{z-t_0}{t-z} \times$$

$$\times \frac{(t-t_0)(z-t_1)}{(t-z)^2} = \frac{[(t-z)+(z-t_0)](z-t_1)}{(t-z)^2} = \frac{(z-t_0)(z-t_1)}{(t-z)^2}.$$

As a result of the last equalities, the expression $R_n(\rho; \varphi; z)$ can be represented as the sum of the following integrals

$$\sum_{\sigma=1}^{n} \int_{\tau_\sigma \tau_{\sigma+1}} \sqrt{\frac{t-a}{t-b}} \left[1 + \frac{z-t_0}{t-z} + \frac{z-t_1}{t-z} + \frac{(z-t_0)(z-t_1)}{(t-z)^2} \right] \qquad (3.2.9)$$

$$r_\sigma(\varphi; t; t_0, t_1) dt$$

Firstly, estimate the sum

$$\sum_{\sigma=1}^{n} \int_{\tau_\sigma \tau_{\sigma+1}} \sqrt{\frac{t-a}{t-b}} \; r_\sigma(\varphi; t; t_0, t_1) dt = \sum_{\sigma=v\pm 1, v} + \sum_{\sigma \neq v\pm 1, v} \qquad (3.2.10)$$

$$\sum_{\sigma=v\pm 1, v} \int_{\tau_\sigma \tau_{\sigma+1}} r_\sigma(\varphi; t; t_0, t_1) dt = O\left(h^{1/2}\right) \max_{\substack{\sigma=v\pm 1, v \\ t \in \tau_\sigma \tau_{\sigma+1}}} |r_\sigma(\varphi; t, t_0, t_1)| =$$

$$= O(h^{1/2})M \quad \left(M = \sup_{t\in\Gamma}|\varphi''(t)| \right),$$

Uniformly with respect to ν, t_0, t_1.

Now, estimate the sum $\sum_{\sigma\neq\nu\pm1,\nu}$ in (3.2.10), first of all, for the term

$$\sum_{\sigma\neq\nu\pm1,\nu} \int_{\tau_\sigma\tau_{\sigma+1}} \sqrt{\frac{t-a}{t-b}}\; \rho_\sigma^*(\varphi; t, t_0, t_1)dt$$

We use the above given representation of ρ_σ^* and estimate each term of the right hand side of $\rho_\sigma^\times$

$$\sum_{\sigma\neq\nu\pm1,\nu} \int_{\tau_\sigma\tau_{\sigma+1}} \sqrt{\frac{t-a}{t-b}}\; dt \int_{\tau_\sigma t} (t-\tau)d\tau \times$$

$$\times \int_{\tau t_0}\left[\frac{1}{(t_1-\tau)^3} - (t_0-u)\frac{(t_0-u)^2+(t_1-\tau)^2}{(t_0-\tau)^3(t_1-\tau)^3}\right]\varphi''(u)du.$$

Due to the fact that $t_0, t_1 \in \tau_\nu\tau_{\nu+1}$ the modulus of the sum does not exceed

$$O(h^{3-1/2})M \max \sum_{j=1}^{n}\frac{1}{j^2h^2} = O(h^{1/2})\,M.$$

Similarly can be estimated the following sum:

$$\sum_{\sigma\neq\nu\pm1,\nu} \int_{\tau_\sigma\tau_{\sigma+1}} \sqrt{\frac{t-a}{t-b}}\,dt \int_{\tau_\sigma t} (t-\tau)d\tau \times$$

$$\times \int_{\tau t_0} (t_0-u)\frac{(t_0-\tau)^2+(t_1-\tau)^2}{(t_0-\tau)^3(t_1-\tau)^3}]\varphi''(u)du.$$

which also gives estimation $O(h^{1/2})M$.

In order to estimate the third sum we take into account that

$$\left|\frac{1}{t_1-t_0}\int_{t_0t_1}(t_1-u)\varphi''(u)du\right| \le c_2|t_1-t_0|M \quad\quad c_2 = const.$$

From here

$$\sum_{\sigma \neq v+1,v} \int_{\tau_\sigma \tau_{\sigma+1}} \sqrt{\frac{t-a}{t-b}} \; dt \int_{\tau_\sigma t} \frac{(t-\tau)d\tau}{(t_1-\tau)^3(t_1-t_0)} \int_{t_0 t_1} (t_1-u)\, \varphi''^{(u)} du$$

Based on the fact that $t_0, t_1 \in \tau_v \tau_{v+1}$ this sum also has an order $O(h^{1/2})M$. Finally we get

$$\left| \sum_{\sigma=1}^{n} \int_{\tau_\sigma \tau_{\sigma+1}} \sqrt{\frac{t-a}{t-b}} \rho_\sigma^*(\varphi; t, t_0, t_1) dt \right| \leq c_3 M h^{1/2} \ln n,$$

where the constants does not depend on t_0, t_1. The expression

$$\sum_{\sigma=1}^{n} \int_{\tau_\sigma \tau_{\sigma+1}} \sqrt{\frac{t-a}{t-b}} l_{\sigma\kappa}(t)\rho_\sigma^*(\varphi; t, t_0, t_1) dt$$

is estimated similarly to the previous reasoning based on the fact that $l_{\sigma\kappa}(t) = O(1) \quad (\kappa = 0,1)$ uniformly with respect to σ and $t \in \tau_\sigma \tau_{\sigma+1}$.

Finally, taking into account the estimation $\rho_\sigma^*(\varphi; t, t_0, t_1)$, we get

$$\sum_{\sigma=1}^{n} \int_{\tau_\sigma \tau_{\sigma+1}} \sqrt{\frac{t-a}{t-b}} r_\sigma(\varphi; t; t_0, t_1) dt = c_1 M h^{1/2} \ln n.$$

As for the remaining sum of the corresponding integrals from (3.2.9), let us assume that t_0, for the given r, is the closest point to z on the line Γ, and let us assume that $t_1 \to t_0$. These estimates can be reduced to estimates of $\sum_{\sigma=1}^{n} \int_{\tau_\sigma \tau_{\sigma+1}} \sqrt{\frac{t-a}{t-b}} z_\sigma(\varphi; t, t_0, t_1) dt$, which can be carried out similarly to the ones discussed earlier. The constants involved in these estimates are independent of the variables t_0, t_1, therefore the estimates hold for any z, if we assume in the computational formulas that $t_1 = t_0$, as mentioned earlier. To calculate $F'''(\rho, \varphi, z)$, we should use three corresponding parameters t_0, t_1, t_2 (finaly here, in the calculating formulas, we also assume that $t_0 = t_1 = t_2$). That is, in this case, we must realize approximations with third order finite differences $\varphi(t, t_0, t_1, t_2)$.

In this case, to obtain the required order of accuracy, instead of two point linear interpolation of the finite differences, we should use quadratic interpolation with three points on the arcs $\tau_\sigma \tau_{\sigma+1}$. At this the function $\varphi(t)$ should have at least third-order bounded derivative.

Under these conditions, the order of convergence $O(h^{1/2})$ is achieved. It should be noted that under such requirements for $\varphi(t)$ and in the case of two-parameter t_0, t_1 formulas, when using quadratic interpolation, it is possible to achieve orders $O(h^{3/2}) \frac{1}{\sqrt{z^2-b^2}}$ and $O(h^{3/2}) \frac{1}{\sqrt{z^2-a^2}}$, depending on the proximity of the point z to either endpoint.

It should be noted that when we take the limit in formula (3.1.3) as $z \to t_0$, we obtain formulas for singular integrals with kernel $(t - t_0)^{-1}$, over the contour $\Gamma \equiv ab$. These formulas coincide with the general expressions for singular integrals provided in §1.1.

If $m = 2$, we can be convinced that in this case, the corresponding estimates for the remainder term do not differ (in terms of the order) from the estimates derived in §1.1. Thus, formula (3.1.3) can be used for calculation of the Cauchy type singular integrals as well as for calculation of (boundary values of the Cauchy type integrals) singular integrals.

Below are tables of approximate evaluating of Cauchy-type integrals for various functions $\varphi(t)$ and $\rho(t)$, for various values of n, at specific points, near the ends.

By $R_n(F)$, the error between exact and approximate values of the Cauchy type integral is denoted.

Table 10 $\qquad \varphi(t) = t^2 \; \rho(t) = \sqrt{\dfrac{1}{(t+1)(t-1)}} \; n = 10$

z	$F(\rho; \varphi, z)$	$F_n(\rho; \varphi, z)$	$R_n(F)$ $= F(\rho; \varphi, z)$ $- F_n(\rho; \varphi, z)$
-1+0.005i	0.5-0.0025i	0.399646-0.0234i	0.10036+0.0209i
1+0.005i	-0.5-0.0025i	-0.399646-0.0234i	-0.10036 + 0.0209i

Table 11 $\varphi(t) = t^2 \; \rho(t) = \sqrt{1/(t+1)(t-1)} \; n = 50$

z	$F(\rho;\varphi,z)$	$F_n(\rho;\varphi,z)$	$0.0266799 + 0.000136i$
$-1 + 0.005i$	$1 + 0.005i$	$0.5 - 0.0025i$	$0.0266799 + 0.000136i$
$1 + 0.005i$	$-0.5 - 0.0025i$	$-0.47332 - 0.0026362i$	$-0.0266799 + 0.000136i$

Table 12 $\varphi(t) = t^2 \; \rho(t) = \sqrt{\dfrac{1}{(t+1)(t-1)}} \; n = 100$

z	$F(\rho;\varphi,z)$	$F_n(\rho;\varphi,z)$	$R_n(F) = F(\rho;\varphi,z) - F_n(\rho;\varphi,z)$
$-1 + 0.005i$	$0.5 - 0.0025i$	$0.487516 + 0.0123104i$	$0.012484 + 0.00981013i$
$1 + 0.005i$	$-0.5 - 0.0025i$	$-0.487516 + 0.0123104i$	$-0.012484 + 0.00981013i$

Table 13 $\varphi(t) = Re(t) + Im(t) \; \rho(t) = \sqrt{\dfrac{t+1}{t-1}} \; n = 10$

z	$F(\rho;\varphi,z)$	$F_n(\rho;\varphi,z)$	$R_n(F) = F(\rho;\varphi,z) - F_n(\rho;\varphi,z)$
$-1 + 0.005i$	$5.84648 \cdot 10^{-15} + 0.0025i$	$-0.00359624 + 0.00348727i$	$0.00359624 - 0.001987266i$
$1 + 0.005i$	$1. + 0.0025i$	$0.866977 - 0.0122782i$	$0.133003 + 0.0147782i$

Table 14 $\varphi(t) = Re(t) + Im(t) \; \rho(t) = \sqrt{\dfrac{t+1}{t-1}} \; n = 50$

z	$F(\rho;\varphi,z)$	$F_n(\rho;\varphi,z)$	$R_n(F) = F(\rho;\varphi,z) - F_n(\rho;\varphi,z)$
$-1 + 0.005i$	$5.84648 \cdot 10^{-15} + 0.0025i$	$-0.00469624 + 0.0003381i$	$0.00469666 - 0.002161i$
$1 + 0.005i$	$1. + 0.0025i$	$0.93816 - 0.032843i$	$0.0618041 + 0.0353434i$

Table 15 $\quad \varphi(t) = Re(t) + Im(t)\; \rho(t) = \sqrt{\frac{t+1}{t-1}}\; n = 100$

z	$F(\rho; \varphi, z)$	$F_n(\rho; \varphi, z)$	$R_n(F)$ $= F(\rho; \varphi, z)$ $- F_n(\rho; \varphi, z)$
$-1 + 0.005i$	$5.84648 \cdot 10^{-15}$ $+ 0.0025i$	-0.0095900942 $+ 0.0039992i$	0.009004 $- 0.005i$
$1 + 0.005i$	$1. + 0.0025i$	0.952251 $- 0.0477457i$	0.0477493 $+ 0.0502457i$

Table 16 $\quad \varphi(t) = Re(t) + Im(t)\; \rho(t) = \sqrt{(t+1)(t-1)}\; n = 10$

z	$F(\rho; \varphi, z)$	$F_n(\rho; \varphi, z)$	$R_n(F)$ $= F(\rho; \varphi, z) - F_n(\rho; \varphi, z)$
-1 $+ 0.005i$	$-0.2499 + 0.005i$	-0.256927 $+ 0.00496i$	$0.00693993 + 0.00050i$
1 $+ 0.005i$	$-0.2499 - 0.005i$	-0.256927 $- 0.00496i$	$0.00693993 - 0.00050i$

Table 17 $\quad \varphi(t) = Re(t) + Im(t)\; \rho(t) = \sqrt{(t+1)(t-1)}\; n = 50$

z	$F(\rho; \varphi, z)$	$F_n(\rho; \varphi, z)$	$R_n(F)$ $= F(\rho; \varphi, z) - F_n(\rho; \varphi, z)$
-1 $+ 0.005i$	-0.249987 $- 0.005i$	-0.256326 $- 0.00155i$	0.009388230 $+ 00680058i$
1 $+ 0.005i$	-0.249987 $- 0.005i$	-0.256326 $- 0.00155i$	0.009388230 $+ 00680058i$

Table 18 $\quad \varphi(t) = Re(t) + Im(t)\; \rho(t) = \sqrt{(t+1)(t-1)}\; n = 100$

z	$F(\rho; \varphi, z)$	$F_n(\rho; \varphi, z)$	$R_n(F)$ $= F(\rho; \varphi, z) - F_n(\rho; \varphi, z)$
-1 $+ 0.005i$	-0.2499 $+ 0.005i$	-0.259376 $- 0.001805i$	$0.009388 + 0.0068005i$
$1 + 0.005$	-0.2499 $- 0.005i$	-0.259376 $+ 0.001805i$	$0.009388 - 0.0068005i$

Chapter IV. Numerical Solutions of Crack Problems Using Singular Integral Equations

The real stiffness of rigid bodies essentially depends on defects of real structures. In real materials, there are always different types of many micro-defects, the development of which is related to the applied loads. As the loads increase, they lead to cracks, which in turn may result in local or global fracture of the body. As expirienced, this phenomenon is specifically typical to failure of structure of strong or quasi-strong deformable rigid bodies.

The study of the stiffness of structural elements and structures with existing cracks is of great interest to many renowned researchers. This topic is critical for understanding the mechanical behavior and integrity of materials under stress.

The theory of brittle fracture in solid bodies is thoroughly examined in several monographs, including works by N. Morozov [72], V. Panasyuk, S. Savruk, A. Datsishin [82], V. Panasyuk [81], F. McClintock, A. Argon [69], T. E. Kobori [34], G. Cherepanov [108], and V. Parton, E. Morozov [84]. Additionally, specific chapters in the monographs of N. Muskhelishvili [78] and L. Sedov [104], as well as articles by J. Si [105], G. Cherepanov [107], R. Bantsuri [110], and others, address related topics in detail.

A comprehensive review of the works on this topic is provided in monograph [82].

In these works, one of the most significant studies involves analyzing stress distribution in bodies during the formation of cracks and cavities. To date, many different types of crack-bearing stressed-deformed bodies have been sufficiently solved within the framework of linear elasticity. Primarily, these solutions concern bodies with a single crack or bodies with cracks arranged in a relatively regular manner. The methods used in these solutions are effective only for specific classes of problems.

In the present chapter of this work, numerical solutions for various crack-related problems are studied using singular integral equations. The foundation of the science of fracture mechanics lies in the study of the stressed and deformed states of crack-containing bodies in the vicinity of crack tips (or other similar types of defects). Additionally, it involves examining the criteria for crack propagation in deformable bodies under external field effects.

Fracture mechanics as a field originates from the work of A. Griffith [111-112]. Subsequently, this theory was developed further by G. Irwin [117], E. Orowan [125], and others. In recent times, the number of significant new studies has increased considerably, with their count growing annually. Notably, substantial progress has been achieved in researching the stress-strain state of plates containing cracks.

In the context of assessing the strength of structures and their elements, one of the fundamental issues in the mechanics of brittle fracture is determining the distribution of stresses along cracks, which is characterized by stress intensity factors.

This research direction represents a further development of the problem, focusing on stress concentration in elastic-deformable bodies, particularly around specific types of concentrated cracks.

The study of stress concentration along cracks began as early as 1909 with Kolosov's work [41], who studied the behavior of cracks with a narrow slit, which could be modeled as ellipsoidal cracks. Later, this research was expanded by England [114], who investigated the behavior of ellipsoidal cracks under any loading. Muskhlishvili [78] further solved the problem of stress of hattered plane. The study of fracture mechanics is of particular interest, focusing on the analysis of stress and strain distribution in the context of crack propagation at endpoints of the crack. These issues, both on plane and other positions of cracks are addressed by several research approaches, which include extensive reviews and bibliographic references, as mentioned in [82].

§4.1 The stressed-deformed state of a flexible plane with a straight crack

As noted in [82], the determination of the stressed-deformed state at any point of a flexible isotropic body is possible by finding the components of the stress vector tensor $\sigma_x, \sigma_y, \tau_{xy}$ (Fig. 1) and the components u, v of the displacement vector.

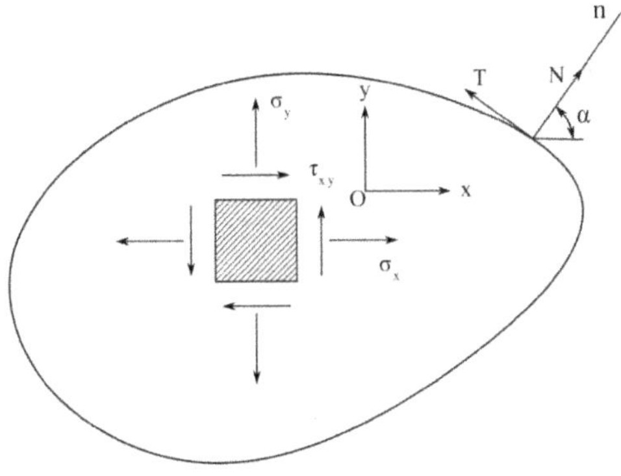

Fig. 1

The statement of the problem and its formulation as an integral equation

Let the area occupied by the elastic body represents the entire plane cut along the segment $|x| \le l, y = 0$ on the axis x. Let us assume that the stress is zero at infinity, while the stress components $\sigma_y^+, \tau_{xy}^+, \sigma_y^-, \tau_{xy}^-$ are given at the edges of the crack.

In the following, the "+" and "–" signs denote the boundary values of the corresponding quantities, respectively, on the upper and lower sides of the crack.

Denote [82]

$$\sigma_y^{\pm}(x, 0) - i\tau_{xy}{}^{\pm}(x, 0) = p(x) \pm q(x), \ |x| < l, \qquad (4.1.1)$$

where $p(x)$ and $q(x)$ are known functions:

$$p(x) = \frac{1}{2}\left(\sigma_y^+ + \sigma_y^-\right) - \frac{i}{2}\left(\tau_{xy}^+ + \tau_{xy}^-\right)$$

$$q(x) = \frac{1}{2}\left(\sigma_y^+ - \sigma_y^-\right) - \frac{i}{2}\left(\tau_{xy}^+ - \tau_{xy}^-\right) \qquad (4.1.2)$$

Let us assume that the discontinuity of the stress and displacement derivatives on the segment $|x| \le l$ is given:

$$\sigma_y^+ - \sigma_y^- - i\left(\tau_{xy}^+ - \tau_{xy}^-\right) = 2q(x) \qquad (4.1.3)$$

139

$$\frac{\partial}{\partial x}[u^+ - u^- + i(v^+ - v^-)] = \frac{i(\aleph + 1)}{2\mu} g'(x) \tag{4.1.4}$$

(where $\aleph$, μ are constants) at this, at the ends of the segment the value of the displacement discontinuity is zero, i.e.

$$g(-l) = g(l) = 0. \tag{4.1.5}$$

Introduce a function

$$\Omega(z) = \overline{\varphi(z)} + z\overline{\varphi'(z)} + \overline{\psi(z)}, \tag{4.1.6}$$

Then displacement and stresses can be expressed by the following relations [41, 78]

$$\sigma_x + \sigma_y = 2\left[\varphi(z) + \overline{\varphi(z)}\right] \tag{4.1.7}$$

$$\sigma_y - i\tau_{xy} = \varphi(z) + \Omega(\bar{z}) + (z - \bar{z})\overline{\varphi'(z)}$$

$$2\mu(u' + iv') = \aleph\varphi(z) - \Omega(\bar{z}) - (z - \bar{z}) - \overline{\varphi'(z)}$$

(4.1.8)

The vanishing in infinity functions $\varphi(z)$ and $\Omega(z)$ are expressed in the following form:

$$\varphi(z) = \frac{1}{2\pi}\int_{-l}^{+l}\frac{Q(t)}{t - z}dt \tag{4.1.9}$$

$$\Omega(z) = \frac{1}{2\pi}\int_{-l}^{-l}\frac{Q(t) + 2iq(t)}{t - z}dt \tag{4.1.10}$$

which solve boundary problems (4.1.3.) and (4.1.4), where $Q(x)$ is defined from the following condition

$$\Phi^+(x) - \Phi^-(x) = i\left[g'(x) - i\frac{2q(x)}{\aleph + 1}\right] = iQ(x) \ |x| < 1 \tag{4.1.11}$$

Taking into account

$$\psi(z) = \overline{\Omega(z)} - \varphi(z) - z\varphi'(z) \qquad (4.1.12)$$

for function $\psi(z)$ we get formula

$$\psi(z) = \frac{1}{2\pi}\int_{-l}^{+l}\left[\frac{\overline{Q(t) - 2iq(t)}}{t - z} - \frac{iQ(t)}{(t - z)^2}\right]dt. \qquad (4.1.13)$$

Let's assume that at the edges of the crack $|x| < l, y = 0$, the load (4.1.1) is given and and there is no stress at infinity. We will find the potentials for this problem $\varphi(z)$ and $\Omega(z)$ in the form of (4.1.9) and (4.1.10). Let us assume that the function $g(x)$ is unknown, and accordingly, the task is to determine either the functions $g(x)$ or $g'(x)$. As it is known, this problem leads to a singular integral equation [82].

$$\frac{1}{\pi}\int_{-l}^{+l}\frac{Q(t) + iq(t)}{t - x}dt = p(x) \quad |x| < l \qquad (4.1.14)$$

from where via condition $\int_{-l}^{l} g'(t)dt = 0$ we obtain

$$g'(x) = -i\frac{\aleph - 1}{\aleph + 1}q(x) + \frac{1}{\pi\sqrt{l^2 - x^2}}\left[-\int_{-l}^{+l}\frac{\sqrt{l^2 - t^2}\,p(t)}{t - x}dt + iR\right],$$

where

$$R = \frac{\aleph - 1}{\aleph + 1}\int_{-l}^{+l}q(t)dt \qquad (4.1.15)$$

For the complete solution of the problem we get

$$\Phi(z) = \frac{1}{2\pi\sqrt{z^2 - l^2}}\int_{-l}^{+l}\frac{\sqrt{l^2 - t^2}\,p(t)dt}{t - z} + \frac{1}{2\pi i}\left[\int_{-l}^{+l}\frac{q(t)dt}{t - z} + \frac{R}{\sqrt{z^2 - l^2}}\right],$$

$$\Omega(z) = \frac{1}{2\pi\sqrt{z^2 - l^2}}\int_{-l}^{+l}\frac{+l\sqrt{l^2 - t^2}\,p(t)dt}{t - z} + \frac{1}{2\pi i}\left[-\int_{-l}^{+l}\frac{q(t)dt}{t - z} + \frac{R}{\sqrt{z^2 - l^2}}\right]. \qquad (4.1.16)$$

Let's consider an infinite isotropic plane, and the stress and displacement distribution in a small neighbourhood of the ends in this plane. To achieve this goal, let's define a new (polar) coordinate system at the tip of the crack $z_{10} = l$ or $z_{20}=-l$ (fig. 2) [82], i.e. assume

141

$$z = \pm(z_1 + l), \quad z_1 = ze^{i\phi} \tag{4.1.17}$$

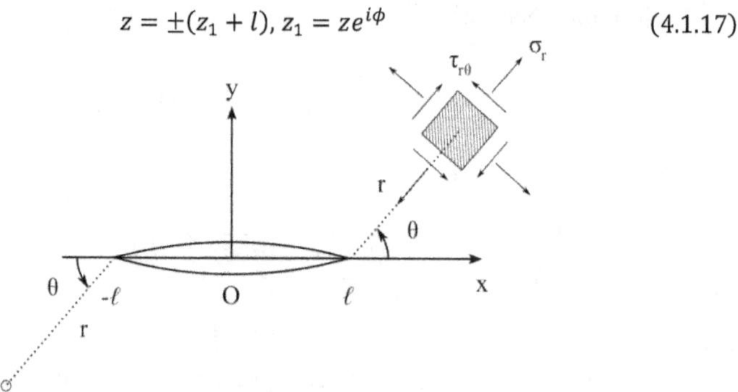

Fig. 2

The potentials $\Phi_1(z_1)$ and $\Omega_1(z_1)$, which correspond (in the new coordinate system) to functions $\Phi(z)$ and $\Omega(z)$ in the system XOY are determined by the formulas

$$\Phi_1(z_1) = \Phi(\pm z_1 \pm l) \tag{4.1.18}$$

$$\Omega_1(z_1) = \Omega(\pm z_1 \pm l)$$

in a small neighborhood of the crack i.e. $|z_1| << l$, the following representation of functions $\Phi_1(z_1)$ and $\Omega_1(z_1)$ is true

$$\Phi_1(z_1) = \frac{k_1^{\pm} - ik_2^{\pm}}{2\sqrt{2z_1}} + O(1)$$

$$\Omega_1(z_1) = \frac{k_1^{\pm} - ik_2^{\pm}}{2\sqrt{2z_1}} + O(1), \tag{4.1.19}$$

where $O(1)$ is bounded value, when $|z_1| \to 0$;

$$k_1^{\pm} - ik_2^{\pm} = -\frac{1}{\pi\sqrt{l}}\left[\int_{-l}^{+l}\sqrt{\frac{l \pm t}{l \mp t}}p(t)dt \pm i\frac{\aleph - 1}{\aleph + 1}\int_{-l}^{+l}q(t)dt\right] \tag{4.1.20}$$

Here and thereafter, $k_1^{\pm}$ and $k_2^{\pm}$ are real quantities. The '+' sign corresponds to the right tip of the crack $(z = l)$, while the lower '-' corresponds to the left tip $(z = -l)$. Coefficients $k_1^{\pm}$ and $k_2^{\pm}$ are named as intensity coefficients of stress. In some works, the intensity coefficients of stress are assigned to $\sqrt{\pi}$ times larger

142

quantities, i.e. $k_1^\pm = \sqrt{\pi}k_1^\pm$, $k_2^\pm = \sqrt{\pi}k_2^\pm$. They are defined by stress functions and parameters, which caracterise the configuration of the body and the shape of the crack and are defined when solving problems of theory of elasticity. In some concrete case formula (4.1.20) has more simple form, e.g.

E.g. when reversed forces P and Q are applied to the point $x = \xi$ $(y = 0)$ of the lower boundary of the crack (fig.3), then (4.1.20)

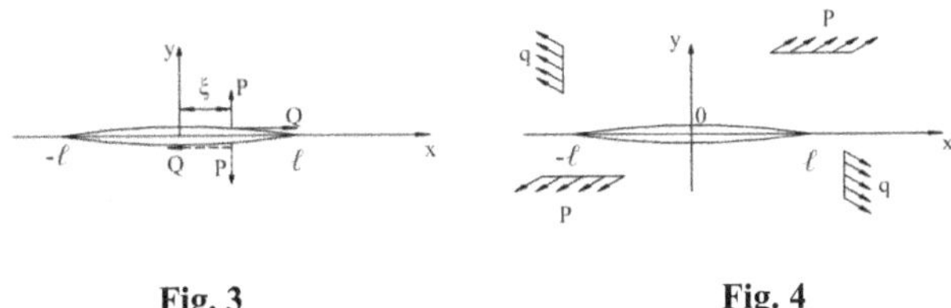

Fig. 3　　　　　　　　　**Fig. 4**

The formula becomes simpler and $k_1^\pm - ik_2^\pm = \dfrac{P-Q}{\pi\sqrt{l}}\sqrt{\dfrac{l\pm\xi}{l\mp\xi}}$ and, when an infinite plane, containing a crack of length $2l$, is subjected to forces p and q acting perpendicular to each other (Fig. 3), equation (4.1.20) becomes even simpler and takes the form

$$k_1^\pm - ik_2^\pm = \tfrac{1}{2}\big[p + q - (p - q)e^{2i\alpha}\big]\sqrt{l}. \qquad (4.1.21),$$

where α is an angle defined by the stress and the crack plane.

During the most general loading along the cracks, the intensity factors $k_1^\pm$ and $k_2^\pm$ are calculated using equation (4.1.20). It should be noted that the same formula can be used to find values of $k_1^\pm, k_2^\pm$ in cases where the loading is applied at the infinity or at an interior point of the plane. For this, it is necessary to determine $\sigma_y(x,0) - i\tau_{xy}(x,0)$, the combination of stresses $|x| < l, y = 0$ on the crack line, on the plane, excluding cut, during the same loading.

Subsequently, the stress intensity coefficients for the plane with cuts are calculated using formula (4.1.20), where it is assumed that $q(x) = 0$; $p(x) = -\big[\sigma_y(x,0) - i\tau_{xy}(x,0)\big]$, in other words, solving such a problem involves applying the method of superposition.

It is significant that the stress and displacement components near the crack tip always have the same functional dependence in polar coordinates (z, Q) [82],

143

corresondingly, the stress intensity factors can be considered as parameters that describe the distribution of stresses resulted by cracks in the material. Therefore, the stress distribution near the crack tip will be known if the stress intensity factors are determined. This is why it is crucial to define the stress intensity coefficients $k_1^{\pm}, k_2^{\pm}$.

Limit Equlibrium Equations

When studying the limit-equilibrium state in stiff solids with crack type defects, we need to determine critical value of external loading at which the cracks begin to propagate, i.e., initiating local or global rupture of the material.

During the solution of this problem, a key moment is the formulation of limit equilibrium conditions.

Such a problem can be easily formulated within the theory of quasi-brittle cracks, when in the neighborhood of the crack tip, the biggest size of the plastic area is small, comparing with the size of the crack and distance from the crack tip to the boundary of the material.

The simplest variant of this problem, based on the physical idea of Griffith [111-112], was developed by Irwin [117].

In the case of normal crack development, i.e., in the case of local symmetry ($k_2 = 0$), he expressed the opinion that in brittle or quasi-brittle materials, crack propagation is initiated when for the given material, under given conditions, the stress intensity factor reaches a certain constant value near the crack-tip.

$$k_1 = \frac{K_{1c}}{\sqrt{\pi}} \tag{4.1.22}$$

The constant K_{1c} which characterizes the material's resistance to fracture, must be determined experimentally.

In general cases, when considering limit equilibrium conditions for material, when the the condition of local symmetry ($k_2 \neq 0$) does not fulfill, additional condition must be introduced.

It is often hypothesized that the initial direction of the crack coincides with a plane where the main part of the tensile stress reaches its maximum value (fig. 2).

The hypothesis is discussed in works [80], [107]. On the ground of it, the angle of the initial propagation is evaluated by formula [81]

$$\theta_* = 2\,arctan\frac{k_1 - \sqrt{k_1^2 + 8k_2^2}}{4k_2}. \tag{4.1.23}$$

The limit equlibrium equation will have the form [81]

$$cos^2\frac{\theta_*}{2}\left(k_1 - 3k_2\,tan\frac{\theta_*}{2}\right) = \frac{K_{1c}}{\sqrt{\pi}}. \tag{4.1.24}$$

From this the critical value of external loading is calculated, when local split of the material begins.

Uniaxial tension of an arbitrarily oriented plate

Let's assume an isotropic thin plate (Fig. 4) with a straight crack of length $2l$ (fig. 4) ($q = 0$) is stretching at the infinity with angle α to the crack plane, under monotonicaly increasing external stress p. At the given moment, the left and right tips of the crack are subjected to identical conditions ($k_{1,2}^{\pm} = k_{1,2}$). Stress intensity coefficients, given by formulas (4.1.21), when $q = 0$, are calculated by formulas

$$k_1 = p\sqrt{l}\,sin^2\alpha\,; \ \ k_2 = p\sqrt{l}\,sin\,\alpha\,cos\,\alpha\,. \tag{4.1.25}$$

The initial propagation angle θ_* of the crack is calculated using the formula

$$\theta_* = 2\,arctan\frac{1 - \sqrt{1 + 8\,cot^2\alpha}}{4\,cot\,\alpha}, \tag{4.1.26}$$

And the limit stress value $p = p_*$ is expressed as a dependency

$$p_* = \frac{p_0}{cos^3\frac{\theta_*}{2}\,sin^2\alpha\,\left(1 - 3ctg\alpha\cdot tg\frac{\theta_*}{2}\right)}, \tag{4.1.27}$$

where

$$p_0 = \frac{K_{1c}}{\sqrt{\pi l}} \qquad (4.1.28)$$

is critical value of external stress at plate tension, which is perpendicular to direction of tension.

Formula (4.1.28) represents the solution of the known problem of Griffith. Using formula (4.1.27), Figure 5 depicts the angle θ_* based on the crack's orientation at a given moment. The initial direction of crack propagation is close to being perpendicular to the direction of the external stress.

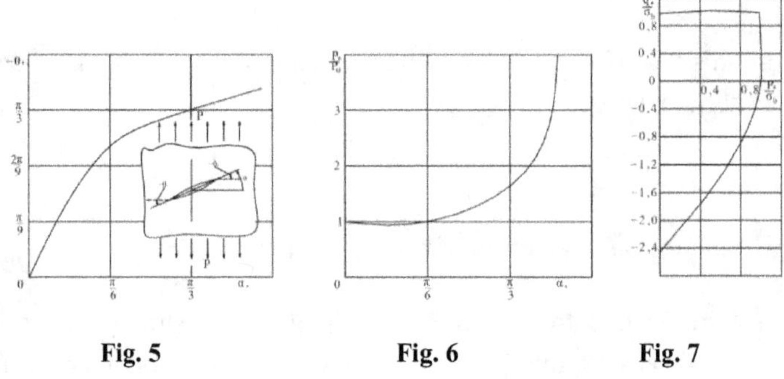

Fig. 5   Fig. 6   Fig. 7

Fig. 6 shows dependence of p_*/p on angle $\alpha\left(0 \leq \alpha \leq \frac{\pi}{2}\right)$. The critical loading p_* attains minimal value $p_* \approx 0{,}97\, p_0$ when $\alpha \approx 1{,}19$ rad.

It should be noted that the dependencies (4.1.26) and (4.1.27), derived based on the hypothesis of the initial crack propagation, align well with experimental data (V. Panasyuk, L. Berezhnitsky, S. Kovchik [80]).

Stress Diagram of Limit Equilibrium

Let us consider an infinite isotropic thin plate of unit width with a straight crack of $2l$, when loadings p and q act orthogonally in the infinity (see Fig. 4), assuming that the crack edges are not in contact, based on formulas (4.1.23) and (4.1.24), for the critical stresses p_* and q_*, we will have:

$$p_* = p_0 \sec^2 \frac{\theta_*}{2} \left[\cos \frac{\theta_*}{2} (\sin^2 \alpha + \eta_0 \cos^2 \alpha) - 3(1 - \eta_0) \sin \alpha \cos \alpha \sin \frac{\theta_*}{2} \right]^{-1}$$

(4.1.28)

$$q_* = \eta_0/p_*$$

here

$$\theta_* = 2 \arctan \frac{1 \mp \sqrt{1 + 8n^2}}{4n}, \tag{4.1.29}$$

where "+" corresponds to $k_1 < 0$ and "-" to $k_1 > 0$;

$$n = \frac{k_2}{k_1} = \frac{(1 - \eta_0) \sin \alpha \cos \alpha}{\sin^2 \alpha + \eta_0 \cos^2 \alpha} \tag{4.1.30}$$

The relationships (4.1.28) - (4.1.29) are used to construct diagram of the critical plain stress for a brittle body weakened with a linear crack.

Suppose we have an initially undeformed elastic body with internal defects whose characterizing linear dimension is $2l$. Assume these defects are isolated and randomly oriented throughout the body. If such a body is subjected to external tensile stress p ($\eta_0 = 0$), the minimum value of the tensile stress is $p_* \approx 0{,}97 p_0$ for different angles α.

In the case under consideration, quantity $\min p_*$ represents the given material σ:

$$\min p_* \approx 0{,}97 p_0 = \sigma \tag{4.1.31}$$

If a body with cracks is subjected to biaxial tension with stresses p and q, then the values of the limit stress p_* and q_* (depending on the fixed values of η_0 the orientation of angle α) are calculated using formulas (4.1.28)–(4.1.30).

In Figure 7, the variation of the limit stress values depending on the parameter η_0 is plotted, where the abscissa represents the values of p_*/σ, and the ordinate represents the values of $\min q_*/\sigma$.

This graph represents the limit stress values p_* сов q_* for cracks, which are valid for crack-hole where the cracks' edges do not interact under the deformation conditions of the body.

In addition to this, it should be noted that min p_*/σ and min q_*/σ, should be considered as the main stress values which are induced in deformed bodies under planar stressed conditions.

As can be seen from the above reasoning, in the analysis of crack propagation, it is essential to calculate stress intensity coefficients, also to determine the limit deformed state of bodies with elastic cracks, after which it is necessary to determine stress redistribution, in order to avoid the failure of the body.

In the mentioned section of the work, numerical calculations of stress intensity coefficients are provided, along with practical methods for solving such problems. In Chapters I-III, we introduced applicable algorithms and provided corresponding numerical tables based on experimental data. The programs are written in symbolic language *Matematica*.

§4.2 Numerical Solution of the Thermo-Insulated Crack Problem

Suppose, on an infinite body along the strip $|x| \leq a, y = 0$, there exists a thermally insulated crack. Let's assume that the temperature distribution without the crack in the body is described by a harmonic function $t_0(x, y)$.

Then the general temperature $T(x, y)$ can be written as follows

$$T(x, y) = t_0(x, y) + t(x, y), \tag{4.2.1}$$

where $t(x, y)$ — is the disturbed temperature field, caused by the crack. Since the crack is thermo-insulated, the condition [82] fulfills at its boudary.

$$\frac{\partial T^+}{\partial y} = \frac{\partial T}{\partial y} = 0, |x| < a, y = 0, \tag{4.2.2}$$

which, via (4.2.1) can be written as:

$$\frac{\partial t^+}{\partial y} = \frac{\partial t^-}{\partial y} = -\frac{\partial t_0}{\partial y} = \varphi(x), |x| < a, y = 0 \tag{4.2.3}$$

148

From the continuity condition of the temperature at the crack tips, it follows that

$$t^+(x,0) = t^-(x,0), \text{ როცა } x = \pm a. \qquad (4.2.4)$$

Thus, the problem is reduced to determining a harmonic function $t(x,y)$ vanishing in infinity that satisfies the conditions (4.2.3) and (4.2.4). Represent temperature $t(x,y)$ in the form

$$t(x,y) = Re\, f\,(z), \qquad (4.2.5)$$

where $f(z)$ is any piecewise holomorphic function. Then

$$\frac{\partial t}{\partial x} = Re\, F\,(z); \frac{\partial t}{\partial y} = -Im\, F\,(z),\ F(z) = f'(z) \qquad (4.2.6)$$

Let us also introduce the redistribution function. As a result, the stress distribution in the vicinity of the crack tip will be known once the stress intensity coefficient is determined. Therefore it is essential to determine intensity coefficients $k_1^{\pm}$ and $k_2^{\pm}$,

$$\psi(x) = \frac{1}{2}[t^+(x,0) - t^-(x,0)],\ |x| < a, \qquad (4.2.7)$$

which describes the jump of temperature $t(x,y)$ at transition via the crack tip. As is known, for $\psi(x)$ the following first-kind singular integral equation is obtained [82]:

$$\frac{1}{\pi}\int_{-a}^{+a}\frac{\psi'(t)dt}{t-x} = \varphi(x),\ |x| < a \qquad (4.2.8)$$

under additional condition

$$\int_{-a}^{+a}\psi'(t)dt = 0 \qquad (4.2.9)$$

The numerical solution of the boundary problem described by equations (4.2.8) - (4.2.9) can be obtained using the algorithms developed in Chapters I-II. Function $F(z)$ has the following form [82].

$$F(z) = \frac{F^*(z)}{\sqrt{z^2 - a^2}} = \frac{1}{\pi i \sqrt{z^2 - a^2}} \int_{-a}^{+a} \frac{\sqrt{a^2 - t^2}\,\varphi(t)dt}{t - z}, \qquad (4.2.10)$$

where

$$F^*(z) = \frac{1}{\pi i} \int_{-a}^{+a} \frac{\sqrt{a^2 - t^2}\,\varphi(t)dt}{t - z}.$$

Knowing the function $F(z)$ from the relationship in equation (4.2.6), we can find the function $f(z)$, and then, using relation (4.2.5), we will determine the temperature $t(x, y)$.

For a fixed point z, we can calculate the integral $F(z)$ using the standard quadratic formula, but the order of accuracy significantly decreases as the point z approaches the boundary points $[-a, +a]$ (especially when $z \to \pm\, a$).

We will calculate the integral $F^*(z)$ using the quadratic formula developed in Chapter III.

$$F^*(z) \approx F_n^*(z) = F^*(1; z)L_v(\varphi; t_0) + p_{v-10}(t_0; z)\frac{\varphi(\tau_{v-1}) - \varphi(t_0)}{\tau_{v-1} - t_0} +$$

$$+[p_{v-11}(t_0, z) + p_{v0}(t_0, z) + p_{v1}(t_0, z) + p_{v+10}(t_0, z)]\frac{\varphi(\tau_{v+1}) - \varphi(\tau_v)}{\tau_{v+1} - \tau_v} +$$

$$+p_{v+11}(t_0, z)\frac{\varphi(\tau_{v+2}) - \varphi(t_0)}{\tau_{v+2} - t_0} + \sum_{\sigma \neq v \pm 1}^{n} \sum_{k=0}^{e} p_{\sigma k}(t_0, z)\frac{\varphi(\tau_{v+1}) - \varphi(t_0)}{\tau_{v+k} - t_0},$$

where

$$p_{\sigma k}(t_0, z) = \frac{1}{\pi i} \int_{\tau_\sigma \tau_{\sigma+1}} \frac{\sqrt{(t - a)(b - t)}(t - t_0)}{t - z} l_{\sigma k}(t)dt$$

$$(\sigma = 1, 2, \cdots, n).$$

Below is a table for the approximate calculation of the integral $F(z)$ using the algorithm mentioned above, for different values n near the point $z = \pm 1$, when $\phi(t) = Re(t) + Im(t)$.

Table 19

n	z	$F(z)$	$F_n(z)$	$R_n(z)$ $= F(z) - F_n(z)$
5	-1 $+0.005i$	3.60173 $+3.46915i$	3.8148 $+3.66829i$	0.21307 $+0.19914i$
	$1+0.005i$	-3.60173 $+3.46915i$	-3.81848 $+3.66829i$	-0.21307 $+0.19914i$
10	-1 $+0.005i$	3.60173 $+3.46915i$	-3.69248 $+3.52953i$	0.09075 $+0.06038i$
	$1+0.005i$	-3.60173 $+3.46915i$	-3.69248 $+3.52953i$	$-0.09075+$ $0.06038i$
30	-1 $+0.005i$	3.60173 $+3.46915i$	3.69248 $+3.52953i$	0.04035 $+0.09338i$
	$1+0.005i$	$-3.60173+$ $3.46915i$	$-3.69248+$ $3.52953i$	-0.04035 $+0.09338i$
50	-1 $+0.005i$	3.60173 $+3.46915i$	3.63742 $+3.45305i$	0.03569 $-0.01610i$
	$1+0.005i$	$-3.60173+$ $3.46915i$	$-3.63742+$ $3.45305i$	-0.03569 $-0.01610i$

Thus, as seen from the table, using the above-mentioned algorithm, the order of accuracy 10^{-2} is achieved for $n = 50$ at the specific points $z = \pm 1 + 0,005i$ (crack endpoints).

§4.3 Numerical Solution of the Crack Problem in a Flexible Body with Longitudinal Displacement

Under longitudinal strain, or anti-planar deformation, stress state is meant in cylindrical bodies, which is caused by the loading along the axis of the cylinder, and which is constant along this axis [82].

If the deformation is oriented along the axis z in a rectangular coordinate system (x, y, z), in static cases the elastic displacements u, v and w in such a body can be represented as follows in the subsequent form [82]

$$u = v = 0, w = w(x, y) \tag{4.3.1}$$

According to Hooke's law, the stress tensor non-zero components τ_{xz} and τ_{yz} are functions of variables x and y, and these components can be expressed via displacement $w(x, y)$ in the following form

$$\tau_{xz} = \mu \frac{\partial w}{\partial x}; \ \tau_{yz} = \mu \frac{\partial w}{\partial y}, \tag{4.3.2}$$

where μ is modulus of shift.

The system of equilibrium equations is reduced to a single equation, which in the presence of mass forces has the following form

$$\frac{\partial \tau_{xz}}{\partial x} + \frac{\partial \tau_{yz}}{\partial y} = \mu \left(\frac{\partial^2 w}{\partial x^2} + \frac{\partial^2 w}{\partial y^2} \right) = 0, \tag{4.3.3}$$

At this, the single equation is identically satisfied in agreement with the deformation.

Since $\mu \, w(x, y)$ represents a harmonic function, let's present it as the real part of some analytical function $f(z)$ of complex variable.

$$w(x, y) = \frac{1}{\mu} Re f(z) \tag{4.3.4}$$

From (4.3.2) it follows

$$\tau_{xz} - i\tau_{yz} = f'(z) = F(z) \tag{4.3.5}$$

where $f(z)$ is an analytical function. The main vector of the force acting on any arc AB occupied by body is calculated (see G.I. Barenblatt, G.I. Cherepanov [5]) by the relationship

$$R = Im[f(z_B) - f(z_A)]. \tag{4.3.6}$$

Let's assume that on an infinite body under longitudinal strain deformation conditions, a cut (crack) is given along the length of the strip $|x| \leq a$, $y = 0$ (Fig. 8).

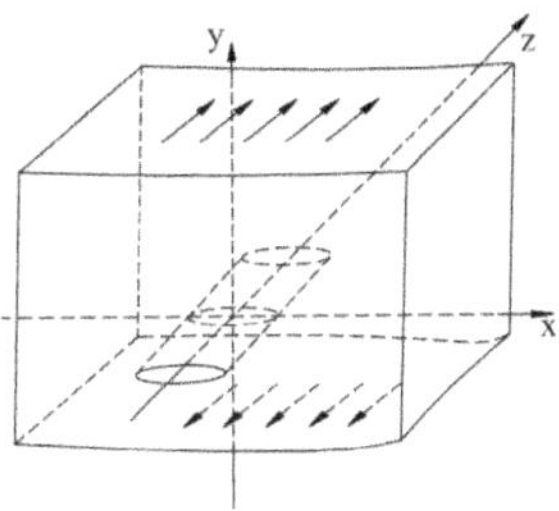

Fig. 8

Let's consider the case where self-equilibrating load acts on the crack surfaces [82].

$$\tau_{yz}^{+} = \tau_{yz}^{-} = \varphi(x), |x| < a \qquad (4.3.7)$$

and the stress at infinity is equal to zero.

Let's denote by $2\mu(x)/\mu$ the displacement discontinuity during the transition along the cross-sectional cut on the plane:

$$w^{+}(x,0) - w^{-}(x,0) = \frac{2\mu(x)}{\mu} \qquad |x| < a. \qquad (4.3.8)$$

It is known that the problem of conjugation of $F(z)$ is obtained [82]

$$F^{+}(x) - F^{-}(x) = 2\mu'(x), |x| < a, \qquad (4.3.9)$$

from where a vanishing (at infinity) piece-wise holomorphic function $F(z)$ is defined by a Cauchy-type integral

$$F(z) = \frac{1}{\pi i} \int_{-a}^{+a} \frac{\mu'(t)dt}{t-x} = \varphi(x), \ |x| < a \qquad (4.3.10)$$

In the case of an unbounded solution at the endpoints $x = \pm a$, an additional condition must be satisfied

$$\int_{-a}^{+a} \mu'(t)dt = 0. \qquad (4.3.11)$$

The approximate solution of the singular integral equation (4.3.10) - (4.3.11) can be found using the algorithms developed in Chapters I and II, from which we can determine the value of the function $\mu'(x)$.

Similarly, for $F(z)$ we have

$$F(z) = \frac{1}{\pi i \sqrt{z^2 - a^2}} \int_{-a}^{+a} \frac{\sqrt{a^2 - t^2}\,\varphi(t)dt}{t - z}. \tag{4.3.12}$$

Now, using formula (4.3.6), we can easily find the displacements of stresses throughout the entire area occupied by the body, specifically in the small vicinity of the crack tip.

The integral (4.3.12) is approximately computed using the algorithm for evaluating Cauchy-type integrals, which we presented in Chapter III. Its effectiveness is demonstrated by the table provided in §4.2.

§4.4 Numerical solution of the problem of two parallel cracks of equal length

Let's assume there are two equal-length cracks on an infinite plate, with their arrangement and the applied load satisfying the symmetry condition (Fig. 9). Then, as is known, this problem is reduced to the following type of singular integral equation [82].

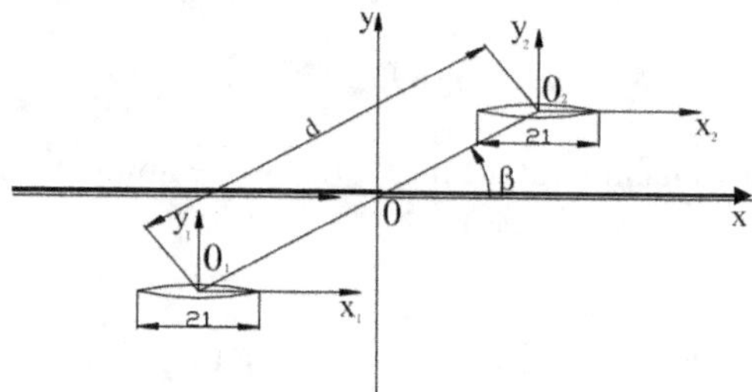

Fig. 9

$$\frac{1}{\pi}\int_{-l}^{+l} \frac{\varphi_0(t)}{t - t_0}\,dt + \frac{1}{\pi}\int_{-l}^{+l} K(t_0, t)\varphi_0(t)dt = f(t_0), \tag{4.4.1}$$

where

$$K(t_0, t) = \frac{t + t_0 + d\cos\beta}{(t + t_0 + d\cos\beta)^2 + d^2 \sin^2\beta}, \tag{4.4.2}$$

where β is the angle between the crack plane and the straight line, which passes through the centers of the cracks.

From equation (4.4.1), in the case of constant loading, the well-known integral equation for stationary (see Smith [121]) or displaced $2l$ ($d\cos\beta = 2l$) parallel cracks (see Okhasi, Isii Kawa, Ekobori [122]) can be easily derived.

Equation (4.4.1) is a first-order singular integral equation. This equation is solved using the algorithms developed in Chapters I and II.

Let's seek the solution

$$\varphi_0(t) = \sqrt{\frac{t+l}{l-t}}\,\varphi(t),$$

where $\varphi(t)$ is a function from Hölder class. For simplicity we put $l = 1$.

$$\frac{1}{\pi}\int_{-1}^{+1}\sqrt{\frac{1+t}{1-t}}\frac{\varphi(t)}{t-t_0}\,dt + \frac{1}{\pi}\int_{-1}^{+1}\sqrt{\frac{1+t}{1-t}}K(t_0,t)\varphi(t)\,dt = f(t_0), \qquad (4.4.3)$$

Let us reduce the integral equation (4.4.3) to its equivalent form.

$$\varphi(t_0) + \frac{1}{\pi}\int_{-1}^{+1}\frac{1}{\pi}\left[\frac{1}{\pi}\int_{-1}^{+1}\sqrt{\frac{1+t}{1-t}}\frac{K(t_1,t)}{t_1-t_0}\,dt_1\right]\varphi_0(t)\,dt = \frac{1}{\pi}\int_{-1}^{+1}\sqrt{\frac{1-t}{1+t}}\frac{f(t)\,dt}{t-t_0} \qquad (4.4.4)$$

In this equation, change the kernel $\dfrac{1}{\pi}\int_{-1}^{+1}\sqrt{\dfrac{1-t}{1+t}}\dfrac{K(t_1,t)}{t_1-t_0}\,dt_1$ by quadrature formula constructed by us in § 1.4

$$\frac{1}{\pi}\int_{-1}^{+1}\sqrt{\frac{1-t}{1+t}}\frac{K(t_1,t)}{t_1-t_0}\,dt_1 \approx L_n[S_n^{(-1/2;1/2)}K),t_0;t], \quad t_0 \in (-1,1),$$

where

$$L_n\left[\left(S_n^{(1/2;-1/2)}K\right);t_0,t\right] = L_{nj}\left[\left(S_n^{(1/2;-1/2)}K\right);t_0,t\right], \quad t_0 \in \tau_j\tau_{j+1}\ (j=\overline{1,n}),$$

$$L_{nj}\left[\left(S_n^{(1/2;-1/2)}K\right);t_0,t\right] = \sum_{k=1}^{n}\frac{\omega_j(t_0)}{(t_0 - t_{jk})\omega_j'(t_{jk})}\left(\left(S_n^{(1/2;-1/2)}K\right);t_{jk},t\right),$$

$$t_0 \in \tau_j \tau_{j+1}$$

Similarly

$$\frac{1}{\pi} \int_{-1}^{+1} \sqrt{\frac{1-t}{1+t}} \frac{f(t)}{t-t_0} dt \approx L_n[S_n^{(-1/2;1/2)} f), t_0], \qquad t_0 \in (-1,1),$$

We will obtain the following system of linear equations

$$\varphi_n(t_{vj}) + \sum_{i=1}^{n} \sum_{e} q_{ie} A_n(t_{ie}, t_{vj}) \varphi_n(t_{ie}) = f_0(t_{vj}), \qquad (4.4.5)$$

where

$$A_n(t_{ie}, t_{vj}) = \left[1 \right.$$

$$+ \sum_{\substack{\sigma=1 \\ \sigma \neq v}}^{n} \sum_{k=1}^{m} \frac{p_{\sigma k}^{*(1/2;-1/2)}}{t_{\sigma \kappa} - t_{vj}}$$

$$\left. + \sum_{\substack{k=1 \\ k \neq j}}^{m} \frac{p_{vk}^{*(1/2;-1/2)}}{t_{vk} - t_{vj}} - p_{vj}^{*(1/2;-1/2)} \sum_{\substack{k=1 \\ k \neq j}}^{m} d_{vk}(t_{vj}) \right] \times$$

$$\times K(t_{vj}; t_{ie}) - \sum_{\substack{\sigma=1 \\ \sigma \neq v}}^{n} \sum_{k=1}^{m} \frac{p_{\sigma k}^{*(1/2;-1/2)}}{t_{\sigma k} - t_{vj}} K(t_{\sigma k}, t_{ie}) - \sum_{\substack{k=1 \\ k \neq j}}^{m} \frac{p_{vk}^{*(1/2;-1/2)}}{t_{vk} - t_{vj}} K(t_{vk}, t_{ie}) +$$

$$+ p_{vj}^{*(1/2;-1/2)} \sum_{\substack{k=1 \\ k \neq j}}^{m} d_{vk}(t_{vj}) K(t_{vk}, t_{ie})]$$

$$f_0(t_{vj}) = [1 + \sum_{\substack{\sigma=1 \\ \sigma \neq v}}^{n} \sum_{k=1}^{m} \frac{p_{\sigma \kappa}^{*(1/2;-1/2)}}{t_{vj} - t_{\sigma \kappa}} + \sum_{\substack{k=1 \\ k \neq j}}^{m} \frac{p_{vk}^{*(1/2;-1/2)}}{t_{vj} - t_{vk}} -$$

$$-p_{vj}^{*(1/2;-1/2)}\sum_{\substack{k=1\\k\neq j}}^{m} d_{vk}(t_{vj})]f(t_{vj}) -$$

$$-\sum_{\substack{\sigma=1\\\sigma\neq v}}^{n}\sum_{k=1}^{m}\frac{p_{\sigma\kappa}^{*(1/2;-1/2)}}{t_{vj}-t_{\sigma k}}f(t_{\sigma k}) - \sum_{\substack{k=1\\k\neq j}}^{m}\frac{p_{vk}^{*(1/2;-1/2)}}{t_{vj}-t_{vk}}f(t_{vk}) +$$

$$+p_{vj}^{*(1/2;-1/2)}\sum_{\substack{k=1\\k\neq j}}^{m} d_{vk}(t_{vk})f(t_{vj})$$

$$(v = 1,2,\cdots,n; \quad j = 1,2\cdots,m)$$

$$p_{\sigma k}^{(1/2;-1/2)} = \frac{1}{\pi i}\int_{\tau_\sigma\tau_{\sigma+1}}\sqrt{\frac{1-t}{t+1}}\prod_{\substack{j=1\\j\neq k}}^{m}\frac{t-t_{\sigma j}}{t_{\sigma\kappa}-t_{\sigma j}}\,dt$$

$$q_{il}^{(-1/2;1/2)} = \frac{1}{\pi i}\int_{\tau_i\tau_{i+1}}\sqrt{\frac{1-t}{t+1}}\prod_{\substack{j=1\\j\neq e}}^{m}\frac{t-t_{ij}}{t_{ie}-t_{ij}}\,dt$$

$$d_{vk} = \frac{\prod_{\substack{j_0=1\\j_0\neq k,j}}^{m}(t_{vj}-t_{vj_0})}{\prod_{\substack{j=1\\j_0\neq k}}^{m}(t_{vk}-t_{vj_0})}.$$

$$p_{\sigma k}^{*(1/2;-1/2)} = \begin{cases} p_{\sigma 1}^{(1/2;-1/2)} + p_{\sigma-1m}^{(1/2;-1/2)} & \sigma - 1,2,3,\dots n;\\ p_{\sigma k}^{(1/2;-1/2)} & k = 2,3,\dots,m-1;\ \ \sigma = 1,2,\dots n \end{cases}$$

$$(\sigma = 1,2,\cdots n; \quad \kappa = 1,2,\cdots m), \quad (t_0 \in \tau_v\tau_{v+1};\ \ v = 1,2,\cdots,n).$$

As we showed in Chapter II, the system (4.4.5) has a unique solution when (4.4.3) has a unique solution, and the accuracy of the approximation is of a sufficiently high order. Below is the solution of this system for several specific cases. The program is written in symbolic language *Matematica*. As shown in the provided tables, the accuracy order 10-3 is achieved for n=10.

$$K(t_0,t) = 1, \qquad f = 2;\ n = 10$$

Table 20. The exact solution of the integral equation is $\varphi(t) = 1$

φ_{00}	0,98437	φ_{21}	0,9955	φ_{42}	0,99942	φ_{63}	0,99695
φ_{01}	0,98641	φ_{22}	0,99734	φ_{43}	0,99891	φ_{70}	0,99695
φ_{02}	0,98811	φ_{23}	0,99769	φ_{50}	0,99891	φ_{71}	0,99856
φ_{03}	0,98955	φ_{30}	0,99769	φ_{51}	0,99077	φ_{72}	0,99879
φ_{10}	0,98955	φ_{31}	0,99802	φ_{52}	0,99184	φ_{73}	0,99899
φ_{11}	0,99	φ_{32}	0,99830	φ_{53}	0,99277	φ_{80}	0,99899
φ_{12}	0,99431	φ_{33}	0,99919	φ_{60}	0,99277	φ_{81}	0,99963
φ_{13}	0,99496	φ_{40}	0,99919	φ_{61}	0,99606	φ_{82}	0,99860
φ_{20}	0,99496	φ_{41}	0,99931	φ_{62}	0,99653	φ_{83}	1,10160

$$k(t_0, t) = (t + t_0 + 4\cos \pi/3)/((t + t_0 + 4\cos \pi/3)^2 + 16\sin^2 \pi/3), f = 2, n = 10$$

Table 21. The exact solution of the integral equation is $\varphi(t) = 2$

φ_{00}	2,03931	φ_{21}	1,95791	φ_{42}	1,86391	φ_{63}	1,93347
φ_{01}	2,03231	φ_{22}	1,92563	φ_{43}	1,85699	φ_{70}	1,93347
φ_{02}	2,0469	φ_{23}	1,91797	φ_{50}	1,85699	φ_{71}	1,89614
φ_{03}	2,01678	φ_{30}	1,91797	φ_{51}	2,00858	φ_{72}	1,88928
φ_{10}	2,01678	φ_{31}	1,91049	φ_{52}	2,00021	φ_{73}	1,88826
φ_{11}	1,98325	φ_{32}	1,90321	φ_{53}	1,99175	φ_{80}	1,88826
φ_{12}	1,97474	φ_{33}	1,87622	φ_{60}	1,99175	φ_{81}	1,85258
φ_{13}	1,96663	φ_{40}	1,87622	φ_{61}	1,94963	φ_{82}	1,84524
φ_{20}	1,96663	φ_{41}	1,86991	φ_{62}	1,941481	φ_{83}	1,87185

§4.5 Numerical solution of the crack problem perpendicular to the boundary of a half-plane

Let's assume that there is a crack of length 1 placed in a flexible half-plane, and the crack is oriented perpendicular to the boundary (as shown in Figure 10).

Assume that a self-balancing load is applied at the edges of the crack, while no stress acts at infinity and along the boundary of the half-plane.

In such a domain, the problem for stress is reduced to solving the following type of singular integral equation [82].

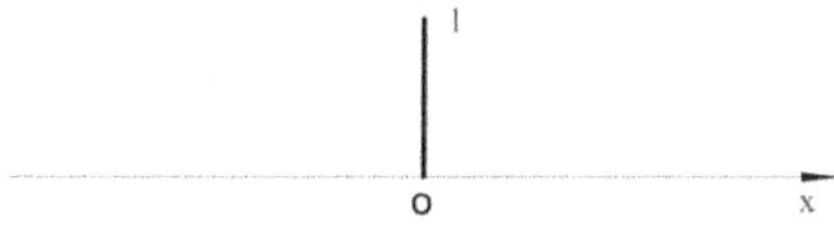

Fig. 10

$$\frac{1}{\pi}\int_{-l}^{+l}\frac{\varphi_0(\tau)}{\tau-y}\,d\tau + \frac{1}{\pi}\int_{-l}^{+l}K(\tau,y)\varphi_0(\tau)d\tau = f(y),\quad |y|<1 \qquad (4.5.1)$$

where

$$K(\tau,y) = \frac{y^2 + 6y + 4\tau y + 2\tau - \tau^2 + 4}{(\tau + y + 2)^3}$$

and $f(y)$ is loading applied on the edges of the crack. Apart from the singularity, $K(\tau,y)$ has a fixed singularity at point $\tau = y = -1$.

This problem has been considered in several works (see e.g., [82], [116]). Below is the numerical solution of this problem, solved with a more simple method. Additionally, the relevant argumentation is provided.

As is known from [32], a fixed singularity can alter the index of the equation, meaning that its solution is also changed. Therefore when such problems are being solved approximately, integrals with fixed singularities should be treated as singular integrals.

In [123], numerical solution algorithms are developed for equations of the form (4.5.1), where integrals with fixed singularities are treated as regular integrals. However, this method is not justified. On the other hand, as noted in [36], only good approximation of singular integrals is insufficient for the solutions of equations to converge to the exact solution of the equation.

A simple solution algorithm for the equation (4.5.1) is developed based on the theory of functional spaces, along with the relevant foundations. The corresponding rstimates are derived. Numerical experiments corresponding to computational calculations are also provided. The program is developed in the symbolic language Mathematica.

159

Represent $\phi_0(\varphi)$ as follows:

$$\varphi_0(\tau) = \frac{\varphi(\tau)}{\sqrt{1-\tau^2}},$$

Then the equation (4.5.1) will have the form:

$$\frac{1}{\pi}\int_{-1}^{+1}\frac{\varphi(\tau)}{\sqrt{1-\tau^2}}\,d\tau + \frac{1}{\pi}\int_{-1}^{+1}K(y,\tau)\frac{\varphi(\tau)}{\sqrt{1-\tau^2}}\,d\tau = f(y), \quad |y| < 1 \qquad (4.5.1')$$

From physical reasoning we get

$$\varphi(-1) = 0 \qquad\qquad\qquad (4.5.2)$$

because, function $\varphi(\tau)$ must have singularity less than $\frac{1}{\sqrt{1+\tau}}$ at the point $\tau = -1$.

The index of equation (4.5. 1'), $\chi = -1$ [77] (according to Muskhelishvili terminology), therefore, for the equation (4.5.1) to have a unique solution $\varphi(\tau)$, the latter must satisfy additional conditions (see [77]).

Now we use the known formula [123]

$$\varphi(\pm 1) = \frac{1}{n}\sum_{k=1}^{n} U_{2n-2}\left[\left(\frac{1\pm\tau_k}{2}\right)^{\frac{1}{2}}\right]\varphi(\tau_k),$$

where

$$U_n(\tau) = \sin[(n+1)\arccos\tau]\,/\,\sin(\arccos\tau);$$

$$T_n(\tau_k) = 0;\ T_n(\tau) = \cos(n\arccos\tau).$$

Thanks to (4.5.2) the additional condition will have form:

$$\frac{1}{n}\sum_{k=1}^{n} U_{2n-2}\left[\left(\frac{1-\tau_k}{2}\right)^{\frac{1}{2}}\right]\varphi(\tau_k) = 0. \qquad\qquad (4.5.3)$$

If we fix the knot $\tau = -1$ (see [44]), then instead of equation equation (4.5.1') we can consider equation

$$\frac{1}{\pi}\int_{-1}^{+1}\sqrt{\frac{1+\tau}{1-\tau}}\frac{\varphi_1(\tau)}{\tau-y}d\tau + \frac{1}{\pi}\int_{-1}^{+1}\sqrt{\frac{1+\tau}{1-\tau}}K(\tau,y)\varphi_1(\tau)d\tau = f(y), \quad |y| < 1 \qquad (4.5.4)$$

whose undex $\chi = 0$. Therefore it will have a unique solution [77]. In otder to base the corresponding method, it is convenient to represent equation (4.5.4) as folows

$$(K\varphi)(y) = (S\varphi)(y) + (H\varphi)(y) = f(y), \qquad (4.5.4')$$

where

$$(S\varphi)(y) = \frac{1}{\pi}\int_{-1}^{1}\frac{(1+\tau)\varphi_1(\tau)d\tau}{\sqrt{1-\tau^2}(\tau-y)},$$

$$(H\varphi)(y) = \frac{1}{\pi}\int_{-1}^{1}\frac{(1+\tau)K(\tau,y)\varphi_1(\tau)d\tau}{\sqrt{1-\tau^2}(\tau-y)},$$

and

$$K(\tau,y) = \frac{y^2 + 6y + 4\tau y + 2\tau - \tau^2 + 4}{(\tau+y+2)^3}.$$

If in (4.5.4') we represent $\rho(\tau) = \frac{1}{\sqrt{1-\tau^2}}$ as a weight function, then for unique solvability, the additional condition (4.5.3) must be used. In such a case, if $\varphi_1(\tau)$ is the solution of equation (4.5.4'), then the relationship $\varphi_1(\tau) = (1+\tau)\varphi(\tau)$ holds. Let us denote (in (4.5.4')) the following:

$$\psi(\tau,y) = \tau - y, \qquad \chi(\tau,y) = (1+\tau)K(\tau,y), \qquad \Omega(\tau,y) = \chi(\tau,y)\psi(\tau,y).$$

It is easy to show that $\chi(\tau,y)$ is a function bounded with respect to both variables on the segment $[-1,1]$. Furthermore, it has such derivatives with respect to both variables (except at $\tau = y = -1$) such that

$$\chi_\tau'|(\tau,y)| < \frac{c_1}{|\tau+y+2|}, \qquad (\tau,y \neq -1), (c_1 = const),$$

$$|\chi_y'(\tau,y)| < \frac{c_2}{|\tau+y+2|}, \qquad (\tau,y \neq -1), (c_2 = const),$$

The function $\psi(\tau, y) = \tau - y$ belongs to the Hölder class with index $\alpha=1$, uniformly with respect to both variables. Following the reasoning presented in [77], it can similarly be shown that the function $\Omega(\tau, y) = \chi(\tau, y)[\psi(\tau, y) - \psi(-1, -1)] = \chi(\tau, y)\psi(\tau, y)$ also satisfies the same property. This property is of critical importance for establishing the method.

In equation (4.5.4), replace the singular integrals with the corresponding quadrature formulas of the highest accuracy

$$(S\varphi)(y_j) \approx (S_n\varphi)(y_j) = \sum_{k=1}^{n} a_{k,n} \frac{\varphi(\tau_k)}{\tau_k - y_j} \tag{4.5.5}$$

$$(H\varphi)(y_j) \approx (H_n\varphi)(y_j) = \sum_{k=1}^{n} a_{k,n} \frac{\Omega(\tau_k, y_j)}{\tau_k - y_j} \varphi(\tau_k), \tag{4.5.6}$$

where

$$\tau_k = \cos\left(\frac{2k-1}{2n+1}\pi\right), \qquad y_j = \cos\left(\frac{2j}{2n+1}\pi\right), \qquad a_{k,n} = 2\pi\frac{1+\tau_k}{2n+1} \tag{4.5.7}$$

$$(k = 1,2, \dots, n; \; j = 1,2, \dots n).$$

Formulas (4.5.5)–(4.5.6) represent high-order quadrature formulas. The equality holds whenever $\varphi(\tau)$ is a polynomial of degree $\leq 2n - 1$. For the equation (4.5.4'), if we interpret the weighting function as $\rho(\tau) = \sqrt{\frac{1+\tau}{1-\tau}}$, then:

$$\tau_k = \cos\left(\frac{2k-1}{2n}\pi\right), \qquad y_j = \cos\left(\frac{j}{n}\pi\right), \qquad a_{k,n} = \pi\frac{1+\tau_k}{n}$$

$$(k = 1,2, \dots, n; \; j = 1,2, \dots n). \tag{4.5.7'}$$

Let's construct a piecewise interpolating function

$$L_n[S_n\varphi, \, y] = L_{nj}[S_n\varphi; \, y], \qquad y \in \tau_j\tau_{j+1} \; (j = 1,2, \dots n - 1),$$

where

$$L_{nj}[S_n\varphi,\ \tau] = \frac{\tau_{j+1} - \tau}{\tau_{j+1} - \tau_j}[S_n\varphi](\tau_j) + \frac{\tau - \tau_j}{\tau_{j+1} - \tau_j}[S_n\varphi](\tau_{j+1}),$$

$$y \in \tau_j\tau_{j+1}\ (j = 1,2,\ldots n - 1).$$

In this context, under the expression $S_n[\varphi;\tau_j]$ $(j = 1,2,\ldots n)$ at the point x0, we understand one of the one-sided limits of $S_n[\varphi;y]$ at the point τ_j. For definiteness, let us assume that $S_n[\varphi;\tau_j] = S_n[\varphi;\tau_j^+]$, $(j = 1,2,\ldots n)$. As is easily seen, for any function φ defined on $[-1,1]$, the operator-function $L_nS_n[\varphi;y]$ is continuous on $[-1,1]$. Furthermore, it satisfies the Hölder condition with an exponent $\alpha = 1$ on this segment.

Analogously to the reasoning presented in §4.4, we can demonstrate that

$$R_n[S_n\varphi,y] = (S\varphi)(y) - L_n[S_n\varphi;y] \in H^{(\alpha-1/2)}[-1,1]\quad \left(\alpha > \frac{1}{2}\right)$$

Analogously to the reasoning presented in §4.4, we can demonstrate that for each $\varphi \in H^{(\alpha)}[-1,\ 1]$ $\left(\frac{1}{2} < \alpha \leq 1\right)$, the following estimate holds:

$$\|R_n[S\varphi,y]\|_{H_\beta} \leq \frac{A\ln n}{n^{\alpha-1/2}}\quad (n > 1, \beta < \alpha - 1/2, \alpha > 1/2)\qquad (4.5.8)$$

Let us consider the equation

$$U_n[L_n[\varphi_n;y] = L_n[S_n(L_n\varphi_n);y] + L_n[H_n(L_n[K_n;y]),y]L_n[\varphi_n,y] = L_n[f;y].$$

(4.5.9)

in the subspace of functions $L_n[\varphi_n;y]$ along with equation (4.5.4)

$$U_n[L_n[\varphi_n;y] = L_n[S_n(L_n\varphi_n);y] + L_n[H_n(L_n[K_n;y]),y]L_n[\varphi_n,y] = L_n[f;y].$$

In this equation, if we assign a value to the parameter y from the set $\{y_j\}$ $(j = 1,2,\ldots n)$, we obtain a system of linear algebraic equations with respect to $\varphi_n(y_j)$

$$[U_n[L_n[\varphi_n;y_j]]]_{y=y_j} = [L_n[f;y_j]]_{y=y_j}\qquad (4.5.9')$$

Starting from a certain $n = n_0$, under above stated conditions, (4.5.9') is uniquely solvable, at this, if $L_n[\varphi; y]$ is a solution to (4.5.9) the following estimation holds:

$$\|\varphi(\tau) - L_n[\varphi_n, \tau]\|_{H_\beta} \leq \frac{c \ln^2 n}{n^{\alpha - \beta - 1/2}}, \qquad (\beta < \alpha - 1/2)$$

where φ is a solution to (4.5.7), and c is some constant.

As an example, the case where a constant normal load $f_k = -1$ is applied on the crack's edge, is considered.

Below is the table of intensity coefficients obtained based on the solution of the system (4.5.9) for various values of n, according to the formulas (4.5.9) and (4.5.7') respectively.

Table 22

n	6	10	20	40	60	80	120
$fk_i/\sqrt{l}$	1,1392	1,1248	1,1224	1,1224	1,1219	1,1217	1,1216
	1,1138	1,1182	1,12064	1,12129	1,12142	1,12146	1,1218

For the problem considered in the previous part, i.e. when

$$k(t, t_0) = k(t_0, t) =$$
$$= \left(t + t_0 + 4\cos\frac{\pi}{3}\right)\Big/\left(\left(t + t_0 + 4\cos\frac{\pi}{3}\right)^2 + 16\sin^2\frac{\pi}{3}\right), f = 2.$$

with the help of above given formulas (4.5.7) and (4.5.7') we have

Table 23

n	6	10	20	40	60	80	120
$fk_i/\sqrt{l}$	1,25457	1,2539	1,25384	1,25298	1,25281	1,25275	1,25279
	1,25362	1,25319	1,25285	1,25279	1,25276	1,25278	1,25277

Below, the graphs (Fig. 11, 12, 13, 14, 15, 16) illustrate the variation of the intensity coefficients' epicurves for different values of $n = 10, 20, 40, 60, 80, 120$.

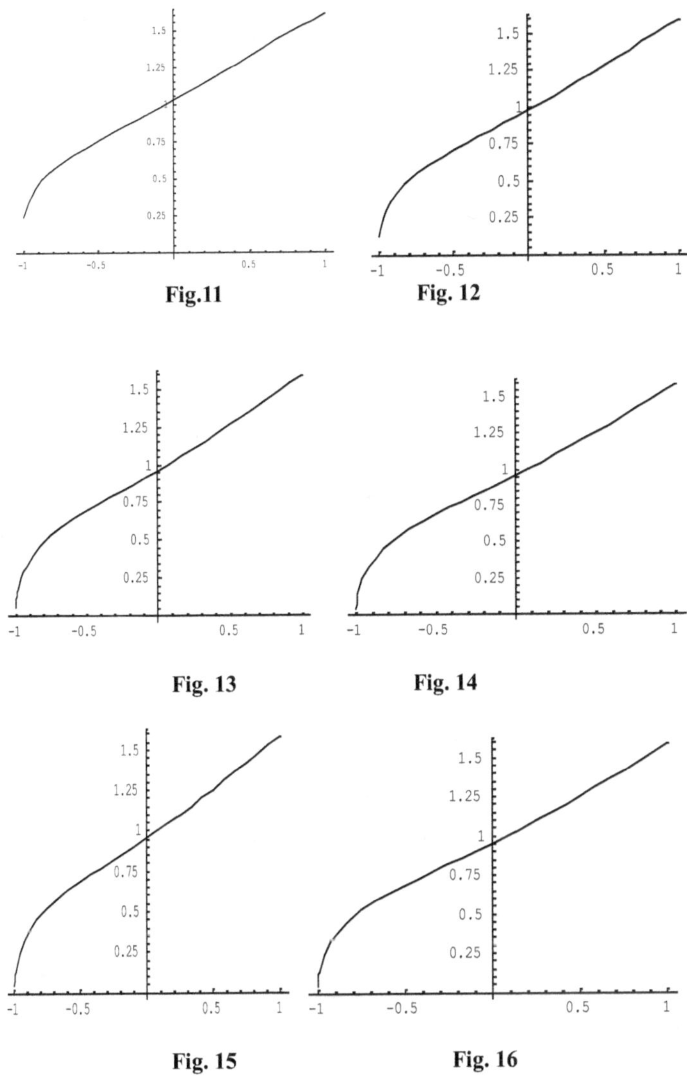

Fig.11   **Fig. 12**

Fig. 13   **Fig. 14**

Fig. 15   **Fig. 16**

Now we will consider the numerical solution of equation (4.5.1) in the case where a point load p is applied at the point $x = \xi$ on the crack's edges (Fig. 17). That is, in the case $p(x) = -p\delta(x - \xi)$, equation (4.5.1) can be written in the following form:

$$\frac{1}{\pi}\int_{-1}^{+1} \varphi_1(\tau)\left[\frac{1}{\tau - y} + K(\tau, y)\right] d\tau = -\frac{2p}{l}\delta(y - \eta), \quad |y| < 1 \qquad (4.5.10)$$

165

where

$$\delta(y, x) = \begin{cases} 1, & x = y \\ 0, & x \neq y. \end{cases}$$

$$\eta = \frac{2\xi}{l} - 1$$

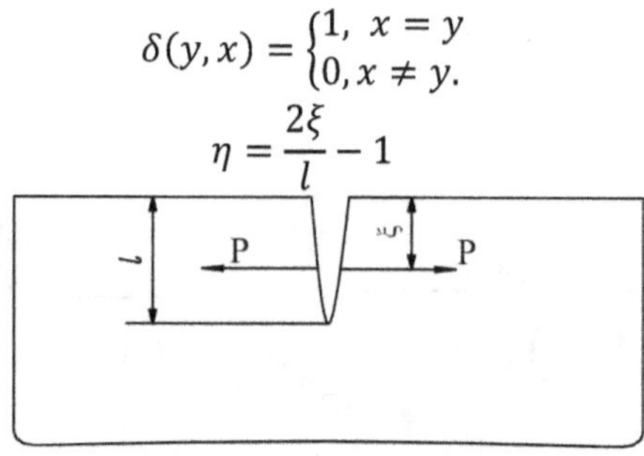

Fig. 17

We will seek the solution to equation (4.5.10) as sum

$$\varphi_1(\tau) = \varphi_0(\tau) + \varphi(\tau) \tag{4.5.11}$$

where function $\phi_0(\tau)$ satisfies the following equation

$$\frac{1}{\pi} \int_{-1}^{+1} \varphi_0(\tau) \left[\frac{1}{\tau - y} + K(\tau, y) \right] d\tau = -\frac{2p}{l} \delta(y - \eta), \quad |y| < 1 \tag{4.5.12}$$

Then, as it is known [82]

$$\varphi_0(y) = \frac{2p}{l\pi\sqrt{1 - y^2}} \frac{\sqrt{1 - \eta^2}}{\eta - y}, |y| < 1 \tag{4.5.13}$$

Therefore, to define the function $\varphi(\tau)$, we will get the equation:

$$\int_{-1}^{+1} \varphi(\tau) \left[\frac{1}{\tau - y} + K(\tau, y) \right] d\tau = -\frac{2p}{l} \sqrt{1 - \eta^2} \int_{-1}^{1} \frac{K(\tau, y) d\tau}{\sqrt{1 - \tau^2}(\eta - \tau)},$$

$$|y| < 1 \tag{4.5.14}$$

In the right hand side of which we have a function, which is bounded for all values of $|y| < 1$ and $|\eta| < 1$, therefore, the numerical solution of equation (4.5.14) can be realized using the method offered above.

§4.6 Crack problems in a plane which are reduced to the first kind singular integral equations

Let us consider the axis Ox in an infinite plane, on which a periodic system of collinear cracks is located (Fig. 18), with the edges subjected to an arbitrary non-self-equilibrating external load.

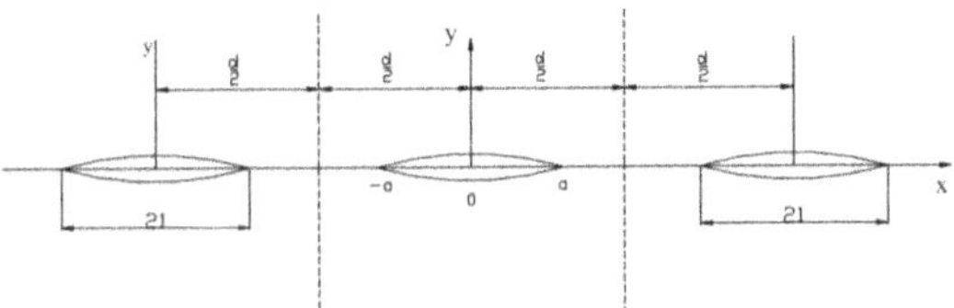

Fig. 18

Then, this crack problem is reduced to the first kind characteristic singular integral equation [82] of the form:

$$\frac{1}{\pi}\int_{-a}^{+a}\frac{R_1(\tau)d\tau}{\tau - y} = -\frac{iR}{d}\cdot\frac{y}{(1+y^2)} + \frac{P_1(y)}{1+y^2}, \qquad |y| < a \qquad (4.6.1)$$

where R is defined from the condition:

$$\int_{-l}^{+l} R(t)dt = i\frac{\aleph - 1}{\aleph + 1}\int_{-l}^{+l} q(t)dt = iR. \qquad (4.6.2)$$

In the case when the infinite plane is weakened by two equal cracks, which are situated simmetrically to an semi-axis (Fig. 19) the following Cauchy kernel singular integral equation is obtained [20]:

$$\frac{1}{\pi}\int_{-c_0}^{+c_0} \phi(\eta)\frac{d\eta}{\eta - y} = p\ (y), \qquad |y| < c_0, \qquad (4.6.3)$$

where

$$P(y) = p(x); \quad c_0 = \frac{1}{2}\left(\cos\frac{2\pi a}{d} - \cos\frac{2\pi b}{d}\right). \qquad (4.6.4)$$

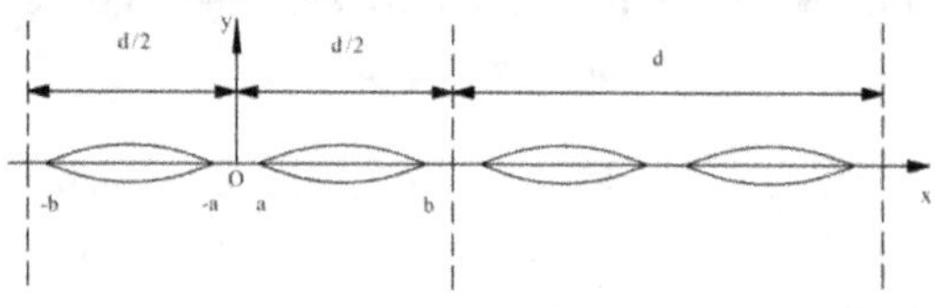

Fig. 19

Now, let us assume that the cracks on the infinite plane are arranged according to Fig. 20. Then, this problem is reduced to equation

$$\frac{1}{\pi}\int_{-a}^{+a}\frac{\varphi(\tau)}{\tau-y}=f(y)+\frac{ABy}{(1-y^2)\pi}\qquad |y|<a\qquad(4.6.5)$$

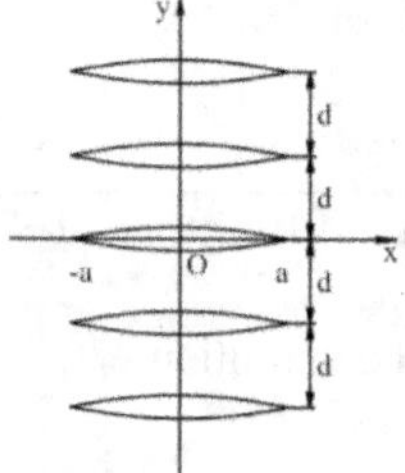

Fig. 20

Let us consider a circular disk of radius R, which has a crack of length $2l$ at its center [82] (Fig. 21).

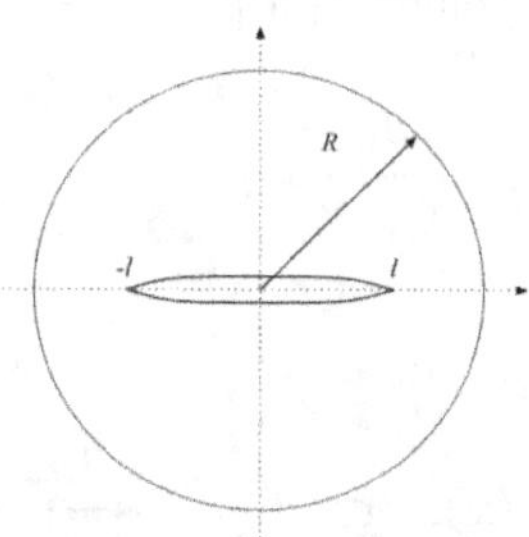

Fig. 21

168

Under symmetric loading along the crack line, the following equation is obtained:

$$\int_{-1}^{1} \left[\frac{1}{\eta - \xi} + M_1(\eta, \xi) \right] v'(\eta) d\eta = \pi\sigma(\xi), \quad |\xi| < 1 \qquad (4.6.6)$$

And under antisymmetric loading:

$$\int_{-1}^{1} \left[\frac{1}{\eta - \xi} + M_2(\eta, \xi) \right] v'(\eta) d\eta = \pi\tau(\xi), \quad |\xi| < 1 \qquad (4.6.7)$$

$$M_1(\eta, \xi) = K(\eta, \xi) + L(\eta, \xi); \quad M_2(\eta, \xi) = K(\eta, \xi) - L(\eta, \xi); \qquad (4.6.8)$$

$$K(\eta, \xi) = \frac{\lambda^2}{2(1 - \lambda^2\eta\xi)^2} [4\xi - 5\eta + \lambda^2\eta(\xi\eta + 3\eta^2 - 3\xi^2) + \lambda^4\xi\eta^2(\xi\eta + \xi^2 - \eta^2) - \lambda^6\eta^4\xi^3];$$

$$L(\eta, \xi) = \frac{\lambda^2}{2(1 - \lambda^2\eta\xi)^2} [2\xi - \eta + \lambda^2\eta(\eta^2 - 2\eta\xi - \xi^2) + \lambda^4\xi\eta^2(\xi\eta + \xi^2 - \eta^2) + \lambda^4\eta^3\xi^2];$$

$$\lambda = \frac{l}{R}, \quad \eta = \frac{t}{l}, \quad \xi = x/l.$$

Let us consider a circular disk of radius R, which has a radial crack of length l on its boundary (fig. 22)

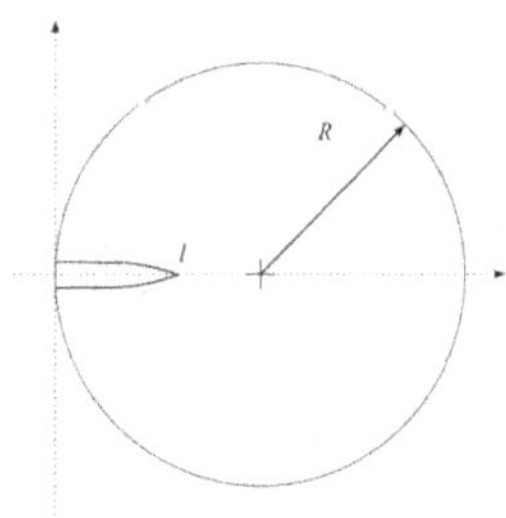

Fig. 22

In the case of symmetric loading on the crack edges, the following equation is obtained:

$$\int_0^1 K(\eta, \xi)\, v'(\eta)\, d\eta = \pi\sigma(\xi), \ 0 \leq \xi \leq 1 \qquad (4.6.9)$$

And under antisymmetric loading:

$$\int_0^1 L(\eta, \xi)\, v'(\eta)\, d\eta = \pi\tau(\xi), \ 0 \leq \xi \leq 1, \qquad (4.6.10)$$

where

$$K(\eta, \xi) = R(\eta, \xi) + S(\eta, \xi), \quad L(\eta, \xi) = R(\eta, \xi) - S(\eta, \xi),$$

$$R(\eta, \xi) = \frac{1}{\eta - \xi} + \frac{1}{2(\eta + \xi - \lambda\eta\xi)^2}[2(\xi^2 + 4\xi\eta - \eta^2) - 2\lambda\eta(3\xi^2 + 7\xi\eta + 2\eta^2) +$$

$$2\lambda^2\eta(\xi^2 + 6\xi^2\eta + 7\xi\eta^2 + \eta^2) - \lambda^2\eta^2\xi(5\xi^2 + 11\xi\eta + 4\eta^2) + \lambda^4\eta^3\xi^2(4\xi + 3\eta) - \lambda^5\eta^4\xi^3]$$

$$S(\eta, \xi) = \frac{\lambda\eta}{2(\eta + \xi - \lambda\eta\xi)^2}[-4\eta + 2\lambda(\xi + \eta)^2 - \lambda^2\eta\xi(3\xi + 2\eta) + \lambda^3\xi^2\eta^2]$$

$$\xi = \frac{x}{l}, \quad \eta = \frac{t}{l}, \quad \lambda = \frac{l}{R}.$$

Equations (4.6.1) – (4.6.10), as we have seen, represent singular integral equations of the first kind, whose numerical solutions are discussed in Chapter II. Since we have already presented the algorithms for solving these equations multiple times, we will not dwell on the numerical solution of these equations here.

In this chapter, we have examined typical crack problems, which are reduced to singular integral equations of the first kind. The wider class of such types of problems have been studied in well-known monographs [16, 66, 72, 77, 82].

Chapter V. Stationary Plane Problems in Aerohydrodynamics

§5.1 The formulation of an aerodynamics problem in the general case

A stationary and non-stationary flow all-around a wing (body) of an arbitrary form, which moves in ideal non-flexible structure with avg forward speed U_0 (fig. 23) is discussed.

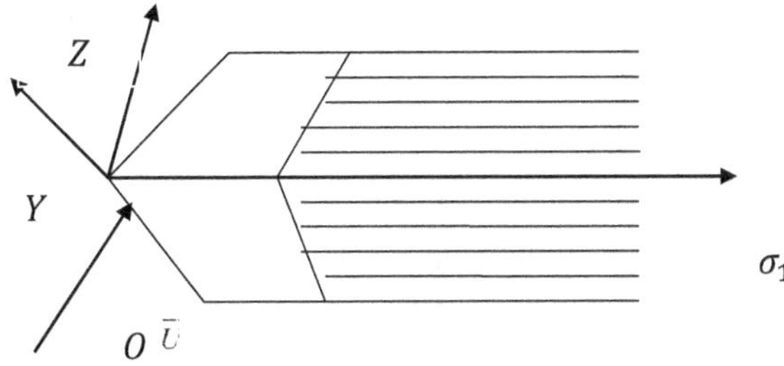

Fig 23

Its density $\rho = \rho_\infty$ does not change during the body's surrounding flow.

In aerodynamics problems, the shape of the body and the law of motion are considered known. If the body is elastic, it is assumed that the law of deformation of the body is also known. Furthermore, naturally, the conditions under which the body moves (or flies) are assumed to be known. Typically, this refers to an infinite environment that is disturbed only by the presence of the body.

On one hand, the motion of the body in a disturbed environment can be considered (such as the effects of wind, flow, turbulence). In this case, the velocity $V(x, y, z, t) = \{V_x, V_y, V_z\}$ and tension $p(x, y, z, t)$ will be unknown in the disturbed environment. We will need Euler's three equations and the continuity equation [27] to determine four unknowns V_x, V_y, V_z and function p.

One of the significant questions is the boundary conditions on the surface of the body. One such boundary condition defines the velocity on the surface of the body. For an ideal fluid, the boundary condition corresponds to the requirement that there be no penetration of the fluid through the body's surface. This is formed as a condittion, that the relative normal component velocity vanishes at the surface of the body.

$$\bar{V}_{rel.}\bar{n}_M = 0, M(x,y,z) = 0,\tag{5.1.1}$$

where $\bar{n}_M$ is a unitary vector of the normal to the surface σ of the body at the considered point $M(x,y,z)$. The further simplification of formulation of the general problem is connected with the next fundamental experienced fact.

During the overflow of the body, an aerodynamic trace will form around it, characterized by the fact that there will be a turbulent flow, i.e. such movement of the fluid particles when they not only move and deformate, but also rotate. The external flow of this environment is not whirling.

Then, the flow around the body and its wake, which is disturbed by the body, can be represented not by three unknown functions $V_x, V_y,$ соs V_z, but by a single velocity potential of field $\Phi(x,y,z,t)$.

At this

$$\nabla\Phi = \frac{\partial\Phi}{\partial x}\bar{i} + \frac{\partial\Phi}{\partial y}\bar{j} + \frac{\partial\Phi}{\partial z}\bar{k} = \bar{V}\tag{5.1.2.}$$

that is

$$V_x = \frac{\partial\Phi}{\partial x}, \quad V_y = \frac{\partial\Phi}{\partial z}, \quad V_z = \frac{\partial\Phi}{\partial z}.$$

In this case, the continuity equation transforms into Laplace's equation

$$\nabla\Phi = \frac{\partial^2\Phi}{\partial x^2} + \frac{\partial^2\Phi}{\partial y^2} + \frac{\partial^2\Phi}{\partial z^2} = 0, \quad M \notin \sigma \cup \sigma_1\tag{5.1.3}$$

where σ_1 represents the wake of the moving body.

Since it is evident physically that the velocity disturbance must dissipate from the body σ to the wake σ_1, i.e., the sought solution $\Phi(x,y,z,t)$ must satisfy the condition:

$$\lim_{r\to\infty}\Phi = 0 \quad r = \sqrt{x^2 + y^2 + z^2}\tag{5.1.4}$$

for the point $M(x, y, z)$ infinitely far from the body σ and its wake σ_1.

Under the above conditions, Euler's equations of motion can be integrated and reduced to Cauchy-Lagrange integral with respect coordinates of p and Φ and time

$$\frac{\partial \Phi}{\partial t} + \frac{\bar{V}_{rel.}}{2} - \frac{\overline{V^{*2}}}{2} + \frac{p - p_\infty}{\rho} = 0, \tag{5.1.5}$$

where $\bar{V}_{rel.}$ is the relative velocity, and $\overline{V^*}$ is the translational velocity of the fluid particles, ρ is the fluid density, and P_∞ is the pressure at infinity, which is considered as known. Thus, the unknown function $p(x, y, z, t)$ can also be eliminated from the general formulation of the problem.

From the general formulation of the problem, one significant peculiarity should be noted. When considering the vortex wake of the body, it is essential to account for the general theorems of aerodynamics, which have not yet been explicitly mentioned in the problem's conditions. These theorems arise from the velocity field and the equations that this field must satisfy. These are:

In steady motion, the vortices are directed along the streamlines.

In unsteady motions, free vortices move along with the fluid particles' trajectories.

The circulation of velocity along any arbitrary closed contour that doesn't intersect neither the body nor its wake remains constant.

Change in the circulation of attached vortices. The requirements stated in point (c) must be fulfilled

Since vortices carry no mass, continuity of pressure must be maintained after passing through the vortices and

$$p^+ = p^-, \quad v^+_{-nm} = v^-_{-nm}, \quad M(x, y, z) \in \sigma_1 \tag{5.1.6}$$

The (+) and (-) indices correspond to different sides of the surface σ_1.

According to Zhukov's theory, the relative velocity of a free vortex is zero, i.e., they move along with the fluid particles. Based on the physical essence of the problem, one important stage in the formulation of the problem is the choice of the overflow scheme. Among these, the following are significant:

a) Non-circulatory flow – This occurs when the vortex wake is typically disregarded during analysis, which is common for highly elongated bodies or during oscillations at the location of the body.

b) The circulatory flow detatched of a vortex, which satisfies all physical apparent conditions, including boundedness of the velocity and pressure across the entire space.

The aforementioned statement affects essentially the choice of the flow scheme.

E.g., when the flow of a thin wing is studied with a rapid flow, it leads to the formation of a specific type of turbulent vortex.

Near and along the entire trailing edge, we apply the Chaplygin-Zhukovsky condition regarding velocity finiteness.

Otherwise, the velocity on them will reach infinitely large values.

It should be noted that even during the ordinary motion of a wing at a constant velocity, a pulsating (non-stationary) flow regime is obtained.

c) The simplified scheme of the circulatory flow around a body, in which certain conditions are thrown off.

More widely spread schemes are those in which the finiteness of velocities and pressures are not required on side sufraces, the leading and trailing edges of the thin wing.

After this, the problem can be solved as stationary, with free vortices generated only on the trailing end of the wing.

§5.2 Airfoil Problems

Let's denote the potential of the disturbed flow velocity by $\Phi = \Phi(x, y)$, while $\bar{V} = grad\ \Phi$ be velocity of this flow at a point on the surface of the airfoil or outside it.

The velocity potential on the airfoil L can be represented in the form of a density $g(M)$ like double potential [66]

$$\Phi(x_0, y_0) = \frac{1}{2\pi} \int_L \frac{\left(\overline{r_{MM_0}}, \overline{n_M}\right)}{r^2_{MM_0}} g(M) ds_M =$$

$$= \frac{1}{2\pi} \int_L \frac{-(x_0 - x)y'_s + (y_0 - y)x'_s}{(x_0 - x)^2 + (y_0 - y)^2} g(M) ds_M \quad M_0 \notin L \qquad (5.2.1)$$

for the points that do not lie on L and

$$\Phi^{\pm}(x_0, y_0) = \frac{1}{2\pi} \int_L \frac{\left(\overline{r_{MM_0}}, \overline{n_M}\right)}{r^2_{MM_0}} g(M) ds_M \pm \frac{1}{2} g(M_0) =$$

$$= \frac{1}{2\pi} \int_L \frac{-(x_0 - x)y'_s + (y_0 - y)x'_s}{(x_0 - x)^2 + (y_0 - y)^2} g(M) ds_M \pm \frac{1}{2} g(M_0), M_0 \notin L \qquad (5.2.2)$$

for the points that lie on L. At this, the sign (+) is taken for the potential value that is obtained when approaching the point M_0 from the side of the airfoil L where the vector $\overline{n_M}$ is directed, while the sign (-) is used for the opposite side.

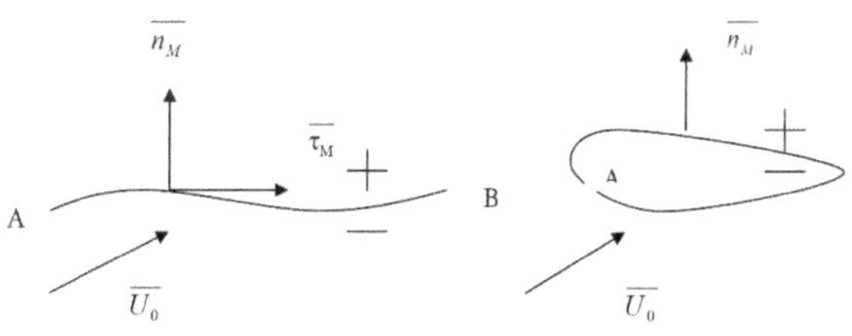

Fig. 24   **Fig. 25**

In formulas (5.2.1) and (5.2.2) and later, in the airfoil problems, we always assume that L is given in parametric form.

The velocity $\bar{V}$ of the disturbed flow is written in the following form:

$$\overline{V} = \frac{1}{2\pi} \int_L \frac{(y_0 - y)\vec{i} - (x_0 - x)\vec{j}}{r_{MM_0}^2} \gamma(M) ds_M, \qquad (5.2.3)$$

for the point M_0 that does not lie on the airfoil L, and

$$\overline{V} = \frac{1}{2\pi} \int_L \frac{(y_0 - y)\vec{i} - (x_0 - x)\vec{j}}{r_{MM_0}^2} \gamma(M) ds_M \pm \frac{1}{2}\left(x'_{0s}\vec{i} + y'_{0s}\vec{j}\right) \qquad (5.2.4)$$

for the points M_0 that lie on the airfoil L.

Thus, as we can see, the velocity field of the potential we have taken is the same as the distribution of the turbulent vorticity intensity $\gamma(M)$ on this layer.

On the other hand, this potential satisfies Laplace equation at every point $M_0 \notin L$, and the velocity tends to zero at infinity.

Thus, the velocity field induced by the vortex, which is distributed along the airfoil L, satisfies the conditions (5.1.2) - (5.1.4).

The boundary condition remains to be satisfied, where the normal component of the relative velocity at the surface of the airfoil is equal to zero.

Since the airfoil is stationary, then

$$\overline{V}_{rel.} = \overline{V} + \overline{U_0} \qquad (5.2.5)$$

and therefore, the problem of finding the disturbed velocity field will be solved if $\gamma(M), M \in L$ satisfies the condition

$$\overline{V_{n_{M_0}}} = -\overline{U}_{\bar{n}_{M_0}}, \qquad M_0 \in L \qquad (5.2.6)$$

i.e.

$$\frac{1}{2\pi} \int_L \frac{y'_{0s}(y_0 - y) + x'_{0s}(x_0 - x)}{r_{MM_0}^2} \gamma(M) ds_M = f(M_0), \quad M_0 \in L,$$

where $f(M_0) = \overline{U}_0 \bar{n}_{M_0}$.

Now let's discuss in more detail the equation (5.2.7) and its solution characteristics for different airfoils.

Let us assume that the contour L is smooth and open, and it is given in parametric form

$$x = x(t),\ y = y(t),\ t \in [-1, +1],\ \text{g.o i.e.}\ \frac{z}{M} = \sqrt{[x'(t)]^2 + [y'(t)]^2}$$

is a continuous function in segment $[-1\ +1]$ and differs from zero.

In this case, the equation (5.2.7) can be written as follows [66]:

$$\frac{1}{2\pi}\int_{-1}^{+1}\frac{k(t_0, t)}{t_0 - t}\gamma(t)dt = f(t_0),\quad t_0 \in [-1, +1]$$

$$k(t_0, t) = \frac{x'(t_0)x_2(t_0, t) + y'(t_0)y_2(t_0, t)}{r_{2,MM_0}^2 r_{MM_0}'}r'_m \qquad (5.2.8)$$

$$x_2(t_0, t) = \frac{x_1(t_0, t)}{t_0 - t},\quad y_2(t_0, t) = \frac{y_1(t_0, t)}{t_0 - t},$$

$$x_1(t_0, t) = x(t_0) - x(t),\quad y_1(t_0, t) = y(t_0) - y(t),$$

$$r_{2,MM_0}^2 = x_2^2(t_0, t) + y_2^2(t_0, t)$$

It is natural to mean that

$$x_2(t_0, t_0) = x'(t_0),\ y_2(t_0, t_0) = x'(t_0)$$

from here $k(t_0, t_0) = 1$ $t_0 \in [-1, 1]$. Now we can represent equation (5.2.8) in the form:

$$\frac{1}{2\pi}\int_{-1}^{+}\frac{\gamma(t)dt}{t_0 - t} + \int_{-1}^{+1}k_1(t_0, t)\gamma(t)dt = f(t_0),\quad t_0 \in [-1, +1]$$

$$k_1(t_0, t) = \frac{1}{2\pi}\frac{k(t_0, t) - 1}{t_0 - t} = \frac{1}{2\pi}\frac{k_2(t_0, t) + k_3(t_0, t)}{r_{2,MM_0}^2 r'}.$$

$$k_2(t_0, t) = [x'(t_0)x_2(t_0, t) + y'(t_0)y_2(t_0, t)]\frac{r_M' - r_{M_0}'}{t_0 - t} \qquad (5.2.9)$$

$$k_3(t_0, t) = r_{M_0}'[x_2(t_0, t)x_3(t_0, t) + y_2(t_0, t)y_3(t_0, t)],$$

$$x_3(t_0, t) = \frac{x(t) - x(t_0) - x'(t_0)(t - t_0)}{(t - t_0)^2},$$

$$y_3(t_0, t) = \frac{y(t) - y(t_0) - y'(t_0)(t - t_0)}{(t - t_0)^2}$$

Starting from the forms of functions $k_2(t_0, t)$ and $k_3(t_0, t)$ we can say that if $x''(t)$, $y''(t) \in H[-1, +1]$, then $k_1(t_0, t)$ also beloings to the class H. Moreover, when $x''(t)$ and $y''(t)$ are small in modulus over the entire segment $[-1, +1]$, then the kernel $k_1(t_0, t)$ is also small in modulus. Therefore, to find the integral characteristics, it is necessary to consider the equation

$$\frac{1}{2\pi} \int_{-1}^{+1} \frac{\gamma(t)dt}{t_0 - t} = f(t_0), \quad t_0 \in [-1, 1] \tag{5.2.10}$$

which is called as a thin, slightly convexed airfoil.

It is known that the equation (5.2.9) and, accordingly, the equation (5.2.7) [77] for a smooth open contour do not have a unique solution.

From the point of view of physics we know that if the point A is the leading edge of the airfoil L (see Fig. 24) surrounding particles that slowly and smoothly flow around the airfoil have to instantaneously turn around the point and the disturbed velocity V must approach infinity as it moves towards this endpoint.

At the endpoint B of the airfoil, the flow should gather rather slowly, therefore, the velocity of the particles, which gather on the upper side of the airfoil L, (as shown in Fig. 24 marked with '+'), should coincide the velocity of the particles gathering on the down side of the airfoil L, more precisely $\bar{V}^+(B) = \bar{V}^-(B)$, this is Zhukovsky-Chaplygin's well-known hypothesis [27].

From the formulas (5.2.4), it directly follows that $\bar{V}^+ - \bar{V}^- = \left(x'_{os}\bar{i} + y'_{os}\bar{j}\right)\gamma(M_0)$, therefore approaching B by L, $\gamma(M_0)$ must tend to zero.

Result 1. During the slow flow around a thin airfoil (circular problem), we should take (5.2.9.), that is, the solution of equation (5.2.7) which equals zero at the point B.

Such solution is unique and γ has the form:

$$\gamma(t) = \sqrt{\frac{1-t}{1+t}}\,\psi(t) \qquad (5.2.11)$$

If a thin airfoil standing on a surface is disturbed by rapidly changing lightweight moving particles of a wind, then we can assume that there is no circulation around the airfoil, meaning that the integral from $\gamma(M)$ along the airfoil is equal to zero [14], [27]. Additionally, both edges are in one and the same condition.

Result 2. In the case of non-circulatory flow around a thin airfoil, the solution $\gamma(M)$ becomes infinite at both ends of the airfoil, i.e., it has the form:

$$\gamma(t) = \frac{\psi(t)}{\sqrt{1-t^2}} \qquad (5.2.12)$$

and satisfies the condition

$$\int_{L} \gamma(M)\,ds = 0 \qquad (5.2.13)$$

Finally, if the flow runs along the airfoil so that it corresponds to a slow flow around the airfoil, with the steady flow occurring along the leading edge, i.e., the flow velocity is finite at the leading edge as well, then such a flow is called non-impact.

Moreover, the flow and the airfoil must be arranged in a specific manner relative to each other. For airplane wings, there is a slat using which the wing airfoil is aligned beneath the flow (the processes of such arrangements are described in [14]).

Result 3. In the case of the non-impact flow around a thin airfoil, the solution $\gamma(M)$ becomes zero across both edges of the airfoil.

$$\gamma(t) = \sqrt{1-t^2}\,\psi(t).$$

§5.3 Numerical Solution of the Problem of a Thin Airfoil in the Case of Circulatory Flow

Let us consider the flow parallel to a plane of an isolated airfoil [66]. Assume that the airfoil itself is stationary.

Under the airfoil, we assume a cylindrical surface, which is formed by the parallel rotation around the axis O_z directed parallel to the plane Oxy (Fig. 25).

Since the parameters of the perturbed flow are not dependent on the parameter z, the airfoil will henceforth always be represented on the plane Oxy solely by the contour L [66]. If L is a simple open contour (Fig. 26), the airfoil is referred to as thin.

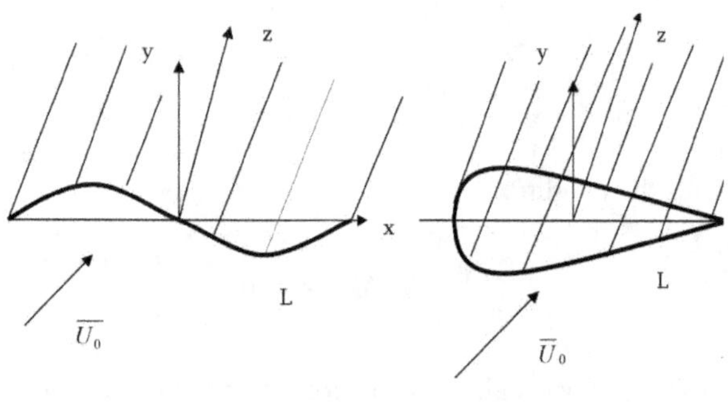

Fig. 26   Fig. 27

If the contour L is simple closed, piece-wise smooth (Fig. 27), then the airfoil is referred to as a spatial.

As is known, during the circulatory flow, regarding the distribution of intensity $\gamma(x)$ along L, for a thin airfoil, the first-order singular integral equation of the following form is obtained.

$$\frac{1}{2\pi} \int_{-1}^{+1} \frac{\gamma(t)}{t_0 - t} dt + \frac{1}{2\pi} \int_{-1}^{+1} k(t_0, t)\gamma(t)dt = f(t_0), \quad t_0 \in [-1, +1], \quad (5.3.1)$$

where $k(t_0, t)$, $f(t)$ are the known functions given on L, and $L = ab$ is a smooth open contour given in parametric form

$$x = x(t), \quad y = y(t), \quad t \in [-1, +1].$$

The numerical solution of this problem in [66] is obtained using the classical method of discrete singularities. However, as noted, this method does not have high accuracy order, and, at this, the estimation of difference between exact and approximate solutions is obtained only in segment $[-1 + \varepsilon, \; 1 + \varepsilon]$ and it is impossible to get estimates at the ends of the integration line with the help of this method. On the other hand, it is especially essential to get information about the solution at these endpoints. Therefore, we solve this problem using the algorithm developed in Chapters I-III, which ensures more accuracy and the corresponding estimation is valid uniformly over the entire closed line, including the ends.

As is known [66], during the slow flow of a thin airfoil L (circular problem), we must take the solution of equation (5.3.1), which becomes zero at the end b of the contour L. This solution is unique and has the following form:

$$\gamma(t) = \sqrt{\frac{1-t}{1+t}}\,\varphi(t), \qquad t \in [-1, +1]$$

where the function $\varphi(t)$ has no singularities on the contour L.

Let's rewrite equation (5.3.1) in its equivalent form

$$\frac{1}{2\pi}\int_{-1}^{+1}\sqrt{\frac{1-t}{1+t}}\frac{\varphi(t)dt}{t-t_0} + \frac{1}{2\pi}\int_{-1}^{+1}\sqrt{\frac{1-t}{1+t}}k(t_0,t)\varphi(t)dt =$$

$$= f(t_0), t_0 \in L \equiv t_0 \in [-1, +1] \tag{5.3.1'}$$

The equation can be derived in the form [26] by inverting the characteristic part.

$$\varphi(t_0) + \frac{1}{2\pi}\int_{-1}^{+1}\sqrt{\frac{1-t}{1+t}}\left[\frac{1}{2\pi}\int_{-1}^{+1}\sqrt{\frac{1+t}{1-t}}\frac{k(t_1,t)}{t_1-t_0}\right]\varphi(t)dt =$$

$$= \frac{1}{2\pi}\int_{-1}^{+1}\sqrt{\frac{1+t}{1-t}}\frac{f(t)}{t-t_0}dt \tag{5.3.2}$$

If we change equation (5.3.2) using the approximation algorithm developed in Chapter II, we will obtain a system of linear equations at the points $\{t_{vj}\}$ $\{v = 1,2,\dots,n \quad j = 1,2,\dots.m\}$ with respect to $\varphi_n(t_{vj})$.

More specifically, this equation will have the following form:

$$\varphi_n(t_{vj}) + \sum_{i=1}^{n}\sum_{s=1}^{m} q_{is}A_n(t_{is},t_{vj})\varphi_n(t_{is}) = g_n(t_{vj}), \qquad (5.3.3)$$

$$A_n(t_{is},t_{vj}) = \{[1 + \sum_{\substack{\sigma=1\\\sigma\neq v}}^{n}\sum_{k=1}^{m}\frac{p_{\sigma k}^*}{t_{\sigma k} - t_{vj}} + \sum_{\substack{k=1\\k\neq j}}^{m}\frac{p_{vk}^*}{t_{vk} - t_{vj}} -$$

$$-p_{vj}^*\sum_{\substack{k=1\\k\neq}}^{m+1} d_{vk}(t_{vj})]k(t_{vj},t_{is}) -$$

$$\sum_{\substack{\sigma=1\\\sigma\neq v}}^{n}\sum_{k=1}^{m}\frac{p_{\sigma k}^*}{t_{\sigma k} - t_{vj}}k(t_{\sigma k},t_{is}) - \sum_{\substack{k=1\\k\neq j}}^{m}\frac{p_{vk}^*}{t_{vk} - t_{vj}}k(t_{vk},t_{is}) +$$

$$+p_{vj}^*\sum_{\substack{k=1\\k\neq j}}^{m+1} d_{vk}(t_{vj})k(t_{vk},t_{is})\}, \qquad (5.3.4)$$

$$g_n(t_{vj}) = \{[1 + \sum_{\substack{\sigma=1\\\sigma\neq v}}^{n}\sum_{k=1}^{m}\frac{p_{\sigma k}^*}{t_{\sigma k} - t_{vj}} + \sum_{\substack{k=1\\k\neq j}}^{m}\frac{p_{vk}^*}{t_{vk} - t_{vj}} - p_{vj}^*\sum_{\substack{k=1\\k\neq j}}^{m+1} d_{vk}(t_{vj})]f(t_{vj}) -$$

$$-\sum_{\substack{\sigma=1\\\sigma\neq v}}^{n}\sum_{k=1}^{m}\frac{p_{\sigma k}^*}{t_{\sigma k} - t_{vj}}f(\sigma_k) - \sum_{\substack{k=1\\k\neq j}}^{m}\frac{p_{vk}^*}{t_{vk} - t_{vj}}f(t_{vk}) + p_{vj}^*\sum_{\substack{k=1\\k\neq j}}^{m+1} d_{vk}(t_{vj})f(t_{vk})\},$$

$$(v = 1,2,\dots.n; \quad j = 1,2\dots.,m).$$

$$p_{\sigma k}^* = \begin{cases} p_{\sigma k}, & k = 2,3,\dots,m-1; \sigma = 1,2,\dots,n;\\ p_{\sigma 1} + p_{\sigma-1m}, & k = 1; \quad \sigma = 1,2\dots,n. \end{cases},$$

$$d_{vk} = \frac{\prod_{\substack{j_0=1\\j_0\neq k,j}}^{m}(t_{vj} - t_{vj})}{\prod_{\substack{j_0=1\\j_0\neq k}}^{m}(t_{vj} - t_{vj})} \qquad (5.3.6)$$

$$p_{\sigma k} = \frac{1}{2\pi}\int_{\tau_\sigma}^{\tau_{\sigma+1}} \sqrt{\frac{1-t}{1+t}}\prod_{\substack{j=1\\j\neq k}}^{m}\frac{t - t_{\sigma j}}{t_{\sigma k} - t_{\sigma j}}\,dt$$

$$q_{is} = \frac{1}{2\pi} \int_{\tau i}^{\tau_{i+1}} \sqrt{\frac{1+t}{1-t}} \prod_{\substack{j=1 \\ j \neq i}}^{m} \frac{t - t_{ij}}{t_{is} - t_{ij}}\, dt.$$

As noted in Chapter II, when $f, k \in H_z(\alpha; L)$ $(1/2 \leq \alpha \leq 1)$ and equation (5.3.1) has a unique solution, then the system (5.3.3) is solvable uniquely and there is the following estimation of the difference between the exact and approximate solutions:

$$O\left(\ln n / n^{r+\alpha+\beta-1/2} \ (1/2 \leq \alpha \leq 1) \right).$$

We conducted a numerical experiment for the thin-airfoil circular flow, which is laid near the Earth [66]. In this case, the airfoil is laid along a straight line $y = 0$, and it is described by the line $y = -H$ (Figure 28).

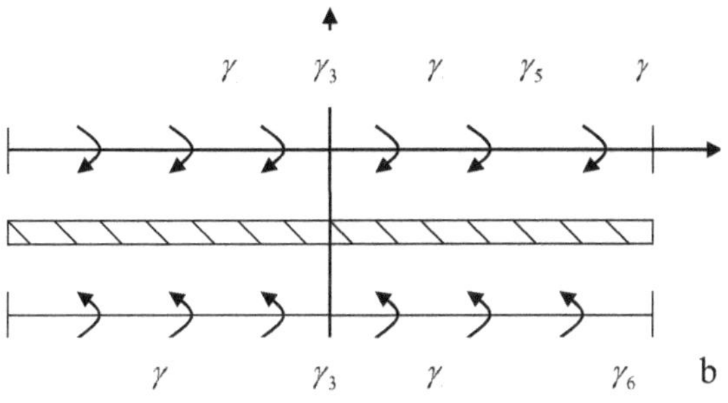

Fig. 28

In this case, the equation for the intensity of the airfoil $\gamma(x)$ is given by

$$\frac{1}{2\pi} \int_{-1}^{+1} \gamma(t) \left[\frac{1}{t - t_0} + \frac{t_0 - t}{(t - t_0)^2 + 4H^2} \right] dt = V^*(t_0),$$

where $V^*(t_0)$ is the speed of the moving flow. For the realization of the algorithm, we have developed a program in symbolic language *Matematica*.

Below (Tables 24, 25, 26) are the values of $\varphi(x)$ based on the following data.

Table 24 $n = 3;$ $m = 2;$ $H = 1;$ $V^* = 2$ (exact solution $\varphi = 2$).

t_{vj}	-1	-0.3333	0.3333	1
$\varphi(t_{vj})$	2.02	2.00	2.00	1.99

Table 25. $n = 3;\quad m = 4;\quad H = 1;\quad V^* = 2$

$t_{v\,j}$	-1	-0.78	-0.56	-0.33	-0.11	-0.11	0.33	0.56	0.78	1
$\varphi(t_{v\,j})$	2.02	2.02	2.01	2.01	2.01	2.00	2.00	1.99	1.99	1.99

Table 26. $n = 5;\quad m = 2;\quad H = 1;\quad V^* = 2$

$t_{v\,j}$	-1	-0.6	-0.2	0.2	0.6	1
$\varphi(t_{v\,j})$	2.02	2.01	2.01	2.00	2.00	1.99

As seen from these tables, the algorithm we developed achieves a sufficiently high accuracy for $n = 3,\ \ n = 5 -$

§5.4 Numerical solution of the thin airfoil problem during circulatory flow with the discrete singularities method of increased order accuracy

As noted in §5.3, the noted problem leads to a first kind singular integral equation

$$\frac{1}{2\pi}\int_{-1}^{+1}\sqrt{\frac{1-t}{1+t}}\frac{\varphi(t)dt}{t-t_0} + \frac{1}{2\pi}\int_{-1}^{+1}\sqrt{\frac{1-t}{1+t}}\,k(t_0,t)\varphi(t)dt = f(t_0) \qquad (5.4.1)$$

where the known functions $K(t_0,t),\ \ f(t_0)$ belong to class H on smooth open contour $L = ab$.

Let us use the fact that $\dfrac{1}{2\pi i}\int\sqrt{\dfrac{1-t}{1+t}}\dfrac{dt}{t-t_0} = \dfrac{1}{2}$ and rewrite the integral equation (5.4.1) as follows

$$\varphi(t_0) + \frac{1}{\pi}\int_{-1}^{+1}\sqrt{\frac{1-t}{1+t}}\frac{\varphi(t)-\varphi(t_0)}{t-t_0} + \frac{1}{\pi}\int_{-1}^{+1}\sqrt{\frac{1-t}{1+t}}k(t_0,t)\varphi(t)dt =$$

$$= f(t_0). \qquad (5.4.1')$$

For any natural n, let us divide the segment $[-1.+1]$ into $2n$ equal parts by points:

$$s_\sigma = -1 + \frac{2}{2n}(\sigma - 1) \qquad (\sigma = 1, 2, \ldots, 2n+1)$$

and denote

$$\tau_\sigma = t(s_\sigma) \quad (\sigma = 1, 2, \ldots, 2n+1).$$

Change

$$\frac{1}{\pi}\int_{-1}^{+1} \sqrt{\frac{1-t}{1+t}}\, \frac{\varphi(t) - \varphi(t_0)}{t - t_0}\, dt$$

by the corresponding (constructed in Chapter II) approximating expression

$$D_n^{(1/2;-1/2)}\left[\left(\overline{S}^{(1/2;-1/2)}\varphi\right); t_0\right]$$

and the regular part

$$\frac{1}{\pi}\int_{-1}^{+1} \sqrt{\frac{1-t}{1+t}}\, k(t_0, t)\, \varphi(t)dt$$

by ordinary quadrature formula. We get the following equation which corresponds to (5.4.1'):

$$\varphi(t_0) + D_n^{(1/2;-1/2)}\left[\left(\overline{S}^{(1/2;-1/2)}\varphi\right); t_0\right] + (K_n\varphi)(t_0) = 2f(t_0) \qquad (5.4.2)$$

In approximate equation (5.4.2), in the first case, we assume that the division points with odd indices $\{\tau_{2\sigma-1}\}$ $(\sigma = 1, 2, \cdots, n+1)$ are taken in the role of interpolational knots, and in the role of parameter t_0 (the so-called calculation points) we take roots of the following polynomial:

$$p_1\prod_{j=2}^{n+1}(t_j - t) + \sum_{\sigma=1}^{n-1}(p_{2\sigma-1} + p1_{2\sigma-1})\left(\prod_{j=1}^{n+1}(\tau_{2j} - t)\right)\Big/(\tau_{2\sigma+1} - t) +$$

$$+p1_{2n-1}\prod_{j=1}^{n}(\tau_{\sigma-2j-1}-t)-\prod_{j=1}^{n+1}(\tau_{2j-1}-t)=0;$$

As experimental calculations have shown, the roots of these polynomials coinside with points

$$\tau_{2j}=t_{0j}=\frac{\tau_{2j-1}+\tau_{2j+1}}{2}\qquad(j=1,2,\dots,n-1)$$

i.e. with division points $\{\tau_\sigma\}$ $(\sigma=1,2,\dots,2n+1)$ with even indices except the last point t_{0n}. The last root t^* of the specified polynomial is close to the point $(t_{0n}=\tau_{2n})$ in the sense of order, namely, $|\tau_{2n+1}-t^*|=O(h/2)$. Based on the above, we can take the point system $\{t_{01},t_{02},t_{03},\cdots,t_{0n-1},t^*\}$ as calculation points t_0.

Then, in the approximate equation (5.4.2), we assume that the points $\{\tau_{2j},t^*\}$ $(j=1,2,\dots,n-1)$ are taken as interpolation nodes, while the calculation points are $\{\tau_{2j-1}\}\{j=1,2,\dots,n+1\}\{\tau_{2j-1}\}\{j=1,2,\dots,n+1\}$. Then, from the approximate equation (5.4.2), we obtain two independent systems of linear equations with respect to the unknowns $\{\varphi(\tau_1),\varphi(\tau_2),\dots,\varphi(\tau_{2n+1})\}$. The resulting system will have the following form:

$$\left[1-\frac{p_1^{(1/2;-1/2)}}{\tau_1-\tau_{2j}}-\sum_{\sigma=1}^{n-1}\frac{\left(p1_{2\sigma-1}^{(1/2;-1/2)}+p_{2\sigma+1}^{(1/2;-1/2)}\right)}{\tau_{2\sigma+1}-\tau_{2j}}-\frac{p1_{2n-1}^{(1/2;-1/2)}}{\tau_{2n+1}-\tau_{2j}}\right]\varphi(\tau_{2j})+$$

$$+\frac{p_1^{(1/2;-1/2)}}{\tau_1-\tau_{2j}}\varphi(\tau_1)+\sum_{\sigma=1}^{n-1}\frac{\left(p1_{2\sigma-1}^{(1/2;-1/2)}+p_{2\sigma+1}^{(1/2;-1/2)}\right)}{\tau_{2\sigma+1}-\tau_{2j}}\varphi(\tau_{2\sigma+1})+$$

$$+\frac{p1_{2n-1}^{(1/2;-1/2)}}{\tau_{2n+1}-\tau_{2j}}\varphi(\tau_{2\sigma+1})=f(\tau_{2j})$$

$$j=\overline{1,n}$$

$$\left[1-\frac{p_2^{(1/2;-1/2)}}{\tau_2-\tau_{2j-1}}-\sum_{\sigma=1}^{n-1}\frac{\left(p1_{2\sigma}^{(1/2;-1/2)}+p_{2\sigma+2}^{(1/2;-1/2)}\right)}{\tau_{2\sigma+2}-\tau_{2j-1}}-\frac{p1_{2n-2}^{(1/2;-1/2)}}{\tau_{2n}-\tau_{2j-1}}\right]\varphi(\tau_{2j-1})+$$

$$+\frac{p_2^{(1/2;-1/2)}}{\tau_2 - \tau_{2j-1}}\varphi(\tau_2) + \sum_{\sigma=1}^{n-1}\frac{\left(p1_{2\sigma}^{(1/2;-1/2)} + p_{2\sigma+2}^{(1/2;-1/2)}\right)}{\tau_{2\sigma+2} - \tau_{2j-1}}\varphi(\tau_{2\sigma+2}) +$$

$$+\frac{p1_{2n-2}^{(1/2;-1/2)}}{\tau_{2n} - \tau_{2j-1}}\varphi(\tau_{2\sigma}) = f(\tau_{2j-1}),$$

$$j = \overline{1, n+1}$$

where

$$p1_{2\sigma-1}^{(1/2;-1/2)} = \frac{1}{\pi i}\int_{\tau_{2\sigma-1}\tau_{2\sigma+1}}\sqrt{\frac{t-a}{t-b}}\frac{(\tau_{2\sigma+1}-t)}{(\tau_{2\sigma+1}-\tau_{2\sigma-1})}dt$$

$$p_{2\sigma+1}^{(1/2;-1/2)} = \frac{1}{\pi i}\int_{\tau_{2\sigma-1}\tau_{2\sigma+1}}\sqrt{\frac{t-a}{t-b}}\frac{t-\tau_{2\sigma-1}}{\tau_{2\sigma+1}-\tau_{2\sigma-1}}dt$$

$$p1_{2\sigma}^{(1/2;-1/2)} = \frac{1}{\pi i}\int_{\tau_{2\sigma}\tau_{2\sigma+2}}\sqrt{\frac{t-a}{t-b}}\frac{(\tau_{2\sigma+2}-t)}{(\tau_{2\sigma+2}-\tau_{2\sigma})}d$$

$$p_{2\sigma+2}^{(1/2;-1/2)} = \frac{1}{\pi i}\int_{\tau_{2\sigma}\tau_{2\sigma+2}}\sqrt{\frac{t-a}{t-b}}\frac{t-\tau_{2\sigma}}{\tau_{2\sigma+2}-\tau_{2\sigma}}dt$$

$$(\sigma = 1, 2 \cdots n - 1).$$

By solving this system, we will obtain the solution of equation (5.4.1) at the points $\{\tau_{2j-1}\}\{j = 1,2,\ldots,2n+1\}$, and then, through interpolation, we can restore the continuous solution.

R e m a r k 1. Let us now assume that the thin airfoil L is piecewise smooth (see Fig. 29).

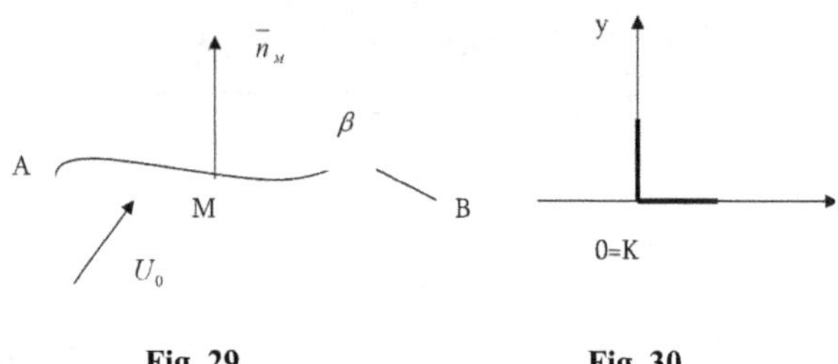

Fig. 29   Fig. 30

The open piecewise smooth airfoil's stationary flow a) with a right angle, b) and let us assume that the point K is one of the corner points. Assume that locally this angle is straight, and the maximal meaning of this angle is β, $\pi < \beta < 2\pi$. In this case, it is known (see [66]) that the discrete intensity $\gamma(M)$ at a point K has a singularity of type $\rho^{\frac{\pi}{\beta}-1}$, where ρ is a distance from the point M to the point K. In this case, in equation (5.4.1), t_0 does not match the values corresponding to the nodes and ends of the line L.

The function $f(t_0)$ has a first-order discontinuity point, while the kernel $K_1(t_0, t)$ in the equation (5.2.9) exhibits a non-integrable singularity at these points. In this case, to solve these equations, we must use the schemes we developed in §1.2-1.3, which take into account the given properties of the functions K_1, f.

R e m a r k 2: Let L be the union of intervals $[0,1]$ on the axes Oy and Ox (see Fig. 30), i.e. $x = 0, y = -t,\ \ t \in [-1, 0]$ and $x = t, y = 0, t \in [0,1]$. Let $\overline{U}_0 = \bar{\imath}$. then $f(t_0) = \overline{U}_0\bar{n}_{Mo} = 1, \qquad t_0 \in (-1,0)$ and $f(t_0) = 0, \qquad t_0 \in (0,1)$.

Let's consider kernels $K(t_0, t)$ and $K_1(t_0, t)$ in the neighbourhood of the origin of the coordinate system. Then, $t_0 \in (1,0)$, and $t \in (0,1)$, we get that

$$K(t_0, t) = \frac{-t_0(t_0 - t)}{t_0^2 + t^2},$$

$$K_1(t_0, t) = \left(\frac{-t_0}{t_0^2 + t^2} - \frac{1}{t_0 - t}\right)\frac{1}{2\pi}.$$

Now it is clear that the function $K(t_0, t)$ is bounded in the vicinity of the point $O(0,0)$, but it is not continuous, and $K_1(t_0, t)$ is not integrable in the vicinity of this point. In this case, it is natural to use the algorithms developed in §1.1 - §1.3.

188

Finally, let's consider the "gate" of the airfoils $[-b, b] \times y_k$ of the system of segments, where $Y_K = kl$, $k = 0, \pm1, \pm2, ..., l$ is a fixed additional number, and $[-b, b]$ is a segment of the axis Ox.

Let's say that the moving flow is not bounded, as during this time the condition of flow for any airfoil is one and the same.

The intensity on all layers of the airfoil depends on coordinate x and does not depend on Y_k, $k = 0, 1, \pm1, \pm2, \pm \cdots$, therefore the condition of leaking should be carried out on a single airfoil, possibly on a airfoil extended along the axis Ox.

If we denote V_k at point M_0 from a k-th airfoil, then the condition of the 0-th airfoil's impermeability will be expressed as:

$$\frac{1}{2\pi} \sum_{k=-\infty}^{\infty} \int_{-b}^{+b} \frac{x_0 - x}{(x_0 - x)^2 + y^2_k} \gamma(x)dx = f(x_0) \qquad x_0 \in (-b, b) \quad (5.4.3)$$

By considering the concept of complex potential, it becomes possible to write:

$$\sum_{k=-\infty}^{+\infty} \frac{1}{2\pi} \frac{x_0 - x}{(x_0 - x)^2 + y_k^2} = \frac{1}{2l} cth \frac{\pi}{l} (x_0 - x) \qquad (5.4.4)$$

Therefore, the equation (5.4.3) can be written as follows:

$$\frac{1}{2\pi} \int_{-b}^{+b} \frac{\gamma(x)dx}{x_0 - x} + \frac{1}{2l} \int_{-b}^{b} K(x_0, x)\gamma(x)dx = f(x_0), \qquad (5.4.5)$$

where

$$K(x_0, x) = \frac{(x_0 - x cth \frac{\pi}{l} (x_0 - x) - \frac{1}{\pi}}{x_0 - x}$$

is an analitic function. For this equation, we can also seek its solutions, which corresponds to the cases of circulatory, non-circulatory, and circumfluence.

We can solve the obtained equation (5.4.5) using the algorithms we developed in Chapters I and II. We have already implemented the realization of this algorithm for similar types of equations, so we will not elaborate on it here anymore.

Chapter VI. On Numerical Solutions of Some Boundary Problems in the Case of Multiply Connected Domains

§6.1 On the Numerical Solution of the Modified Dirichlet Problem for Open Contours

As is known [78], the problem of finding the conformal mapping function for cracks, torsion (to which, in turn, lead many applied problems [78]) is reduced to a modified Dirichlet problem, formulated as follows:

Let L be a smooth, simple, open Liapunov contour. Denote by D the plane cut along the line L. The problem consists in:

We need to find a vanishing at infinity holomorphic function Φ in domain D which is continuously extendable from right and left sides along line L, except the endpoints a and b, where it satisfies the following conditions:

$$|\Phi(z)| < \frac{const}{|z - c|^\alpha} \qquad 0 < \alpha \le const < 1 \qquad (6.1.1)$$

c coinsides with a or b under the following condition

$$Re\, \Phi^+(t_0) = Re\, \Phi^-(t_0) = f(t_0) + d, \qquad (6.1.2)$$

where $f(t_0)$ is a real continuous function given on L and d is determined during the process of solving the problem.

Due to [78] we put

$$U(x,y) + iV(x,y) = \Phi(z) = \frac{1}{\pi}\int_L \frac{\varphi_0(t)dt}{t - z}, \qquad (z \in D)$$

Let us seek the solution in the form

$$U(x,y) = Re\frac{1}{\pi}\int_L \frac{\varphi_0(t)dt}{t - t_0}dt \qquad (z \in D) \qquad (6.1.3)$$

Where $\varphi_0(t)$ is a real function sought. Then the problem of finding the function φ is reduced to a singular integral equation of the following type

$$Re\frac{1}{\pi}\int_L \frac{\varphi_0(t)dt}{t-z} - \frac{1}{\pi}\int_L \varphi_0(t)ds = f(t_0) \qquad (6.1.4)$$

(s is a circular abscissa of the line L).

At this, the positive direction along the contour L is understood as the movement along L in the direction of increase of the parameter s.

Based on the reasoning presented in [78], the equation (6.1.4) is uniquely solvable, and if $\varphi(t)$ is its solution, then the sought solution $u(x,y)$ is given by (6.1.3).

At this, the sought constant involved in boundary condition

$$d = \frac{1}{\pi}\int_L \varphi(t)ds.$$

The equation (6.1.4) can be presented as [78]:

$$\frac{1}{\pi}\int_L \frac{\varphi_0(t)dt}{t-t_0} - i\,Re\frac{1}{\pi i}\int_L \frac{\varphi_0(t)dt}{t-t_0} - \frac{1}{\pi}\int_L \varphi_0(t)ds = f(t_0)$$

or

$$\frac{1}{\pi}\int_L \frac{\varphi_0(t)}{t-t_0}dt + \frac{1}{\pi i}\int_L \varphi_0(t)\frac{\partial\vartheta}{\partial s}ds - \frac{1}{\pi}\int_L \varphi(t)ds = f(t_0), \qquad (6.1.5)$$

where

$$\vartheta = \arg(t-t_0), \qquad \frac{\partial\vartheta(t_0,t)}{\partial s} \in H(L \times L)$$

$$\frac{\partial\vartheta}{\partial s} = \frac{y'(s)[x'(s)-x(s_0)] - x'(s)[y'(s)-y(s_0)]}{r^2} =$$

$$= \frac{y'(s)\cos\vartheta - x'(s)\sin\vartheta}{r} = \frac{y'(s)\cos\vartheta - x'(s)\sin\vartheta}{t-t_0}e^{i\vartheta}$$

where $t = t(s) = x(s) + iy(s)$ is equation of the line L, $r = |t-t_0|$. On this base, the equation (6.1.5) can be written as:

$$\frac{1}{\pi i}\int_L \frac{\varphi_0(t)dt}{t-t_0} + \int_L \frac{K(t_0,t)\varphi_0(t)}{t-t_0}\,dt = f(t_0), \qquad (6.1.6)$$

where the kernel

$$\frac{K(t_0,t)}{t_0-t} \qquad (K(t_0,t) \in H, \quad K(t_0,t_0) \equiv 0));$$

$$K(t_0,t) = \frac{1}{\pi}\{[y'(s)\cos\vartheta - x'(s)\sin\vartheta]e^{i\vartheta} - i(t-t_0)\}t'(s),$$

$$\overline{t'(s)} = \frac{d}{ds}\overline{t(s)}.$$

Additionally, let's take into account that $x'(s) = \cos\vartheta$, $y'(s) = \sin\vartheta$, where ϑ is the angle between the tangent passing through the point $x(s) + iy(s)$ of the line L and the axis Ox. It is easy to verify that $K(t_0,t_0) = 0$. On the other hand, in specific, practical, and applicational problems, in many cases, representing the equation of the line L through circular abscissas is difficult. Therefore, in such cases, let us transform the expression $K(t_0,t)$ so that it corresponds to representation of contour L with any parameter σ; $t = t(\sigma)$.

As direct calculations show

$$y'(s)\cos\vartheta - x'(s)\sin\vartheta = [y'(\sigma)\cos\vartheta - x'(\sigma)\sin\vartheta]\frac{d\sigma}{ds},$$

where, taking into account the notations,

$$x'(\sigma) = \frac{dx}{d\sigma}, \quad y'(\sigma) = \frac{dy}{d\sigma}.$$

besides, due to

$$\overline{t'(s)} = \frac{d\bar{t}}{d\sigma}\cdot\frac{d\sigma}{ds} = \frac{1}{|t'(\sigma)|}\overline{t'(\sigma)},$$

we have

$$\frac{d\sigma}{ds} = \frac{\overline{t'(\sigma)}}{|t'(s)|^2}$$

and we can write

$$K(t_0, t) = \frac{1}{\pi i} \left\{ \frac{[y'(\sigma) \cos \vartheta - x'(\sigma) \sin \vartheta] e^{i\vartheta}}{|t'(\sigma)|} - i(t - t_0) \right\} \frac{\overline{t'(\sigma)}}{|t'(\sigma)|}.$$

This can be modified in a certain way.

Noting (based on the reasoning provided in [78]) that in the context of solving the equation (6.1.6), we can pass from the integral $\int_L \varphi(t) ds$ to an integral

$$\int_L \varphi(t) d\sigma = \int_L \varphi(t) \frac{dt}{t'(\sigma)}.$$

Finally, taking this into account, the initial problem is reduced to the following singular integral equation of the first kind

$$\frac{1}{\pi} \int_L \frac{\varphi_0(t) dt}{t - t_0} + \frac{1}{\pi i} \int_L \frac{[y'(\sigma) \cos \vartheta - x'(\sigma) \sin \vartheta] e^{i\vartheta} - i(t - t_0)}{t - t_0} \cdot \frac{\varphi_0(t) dt}{t'(\sigma)} =$$

$$= f(t_0) \tag{6.1.7}$$

After finding real part of $\varphi(t)$, we can express the desired harmonic function $U(x, y)$ based on the above reasoning, and for the constant d we have:

$$d = \frac{1}{\pi} \int_L \varphi_0(t) \frac{dt}{t'(\sigma)}.$$

Let us use the known representations of solutions of the first-kind singular integral equation and the reasoning presented in Chapter II. Then, if we seek the solution $\varphi_0(t)$ in the form $\varphi_0(t) = \sqrt{(t-a)/(t-b)}\varphi(t)$, where $\varphi \in H$, then equation (6.1.7) will take the form:

$$\frac{1}{\pi} \int_L \sqrt{\frac{t-a}{t-b}} \frac{\varphi(t) dt}{t - t_0} + \frac{1}{\pi i} \int_L \sqrt{\frac{t-a}{t-b}} K_1(t, t_0) \varphi(t) dt = f(t_0) \tag{6.1.8}$$

$$K_1(t_0, t) = \frac{[y'(\sigma) \cos \vartheta - x'(\sigma) \sin \vartheta]}{t - t_0} \cdot \frac{1}{t'(\sigma)}.$$

As is well known, [77], the equation has a unique solution of the class H.

When we seek the solution in the form $\varphi_0(t) = \dfrac{\varphi(t)}{\sqrt{(t-a)(t-b)}}$ the equation (6.1.7) will have the form

$$\frac{1}{\pi}\int_L \frac{\varphi(t)dt}{\sqrt{(t-a)(t-b)(t-t_0)}} + \frac{1}{\pi i}\int_L \frac{K_1(t_0,t)\varphi(t)dt}{\sqrt{(t-a)(t-b)}} = f(t_0), \qquad (6.1.9)$$

In such a case, for the unique solution of equation (6.1.9), the solution must satisfy the additional condition

$$\int_L \frac{\varphi(t)dt}{\sqrt{(t-a)(t-b)}} = C,$$

where C is a constant.

Finally, when we seek the solution in the form $\varphi(t) = \sqrt{(t-a)(t-b)}\,\varphi_0(t)$, the equation (6.1.7) will have the form:

$$\frac{1}{\pi}\int_L \sqrt{(t-a)(t-b)}\,\frac{\varphi(t)dt}{t-t_0} + \frac{1}{\pi i}\int_L \sqrt{(t-a)(t-b)}\,K_1(t_0,t)\varphi(t)dt =$$

$$= f(t_0) \qquad\qquad (6.1.10)$$

For equation (6.1.10) to have a solution, an additional condition $\int_L \dfrac{f(t)dt}{\sqrt{(t-a)(t-b)}} = 0$ must be satisfied. In this case, it will have a unique solution [77].

We will not dwell further on the numerical solutions of equations (6.1.7), (6.1.9), and (6.1.10) here, as this issue was discussed in detail in Chapter II.

Let us use the approximation solution schemes for integral equations outlined in Chapter II for the equations mentioned above. Once we obtain the values of $\{\varphi_j\}_{j=1}^n$ from the corresponding approximate system, we take

$$U_n(x,y) = Re\,\frac{1}{\pi}\int_{ab} \rho(t)\frac{\psi_n(t)}{t-z}dt \quad (z \in D), \qquad (6.1.11)$$

as an approximate solution of the problem where $\psi_n(t)$ represents the Lagrange interpolational polinomial constructed by values $\{\varphi_j\}_{j=1}^n$, and $\rho(t) =$

$(t - a)^\alpha (t - b)^\beta$ is a weight function. α and β take on values depending on which equation's solution we discuss. Specifically, for equation (6.1.8), $\alpha = \frac{1}{2}, \beta = -\frac{1}{2}$; in the case of equation (6.1.9), $\alpha = \beta = -\frac{1}{2}$; and for equation (6.1.10) $\alpha = \beta = \frac{1}{2}$;

Using the formula (6.1.11), we obtain an approximate value $U_n(x,y)$ of the solution for the problem in the domain $D + L$, in addition, the approximate value of the constant d is necessary to determine

$$d_n = \frac{1}{\pi} \int_L \frac{\varphi_n(t)}{t'(s)} dt.$$

Here, $U_n(x,y)$ is the solution to the Dirichlet modified problem. The approximate solution to the Dirichlet problem itself will be $U_n(x,y) - a_n$. In formula (6.1.11), let us use the approximate formula constructed in Chapter III for the points $z = x + iy$ which are close to line L

$$\frac{1}{\pi} \int_L \rho(t) \frac{\psi_n(t)dt}{t - z} \approx 2iL_v(\varphi_v; t_0) + 2i \sum_{\sigma=1}^{n} \sum_{k=0}^{1} p_{\sigma k}(\rho, t_0, z) \cdot \frac{\psi_v(\tau_{\sigma+k}) - L_{\sigma k}}{\tau_{\sigma+k} - z},$$

where the coefficients $p_{\sigma k}(\rho; t_0; z)$ are defined according §3.1. For the rest points z we use ordinary quadratic formulas

$$\frac{1}{\pi} \int_L \rho(t) \frac{\psi_n(t)}{t - z} dt \approx 2i \sum_{\sigma=1}^{n} \sum_{k=0}^{1} p_{\sigma k} \frac{\psi_n(\tau_{\sigma+k})}{\tau_{\sigma k} - z};$$

Below are the tables for the numerical solutions of Dirichlet's problem, where $L \equiv [-1, +1]$ is a segment of the real axis, and $f(t) = f(t(s)) = Re(t) = t$. $U(x,y)$ фо $U_n(x,y)$ represent the exact and approximate solutions of the corresponding problem, respectively. The point z is taken from the vicinity of the ends of the contour.

Table 27

n	z	$U(x,y)$	$U_n(x,y)$	$R_n = \lvert U - U_n\rvert$
10	$1 + 0.005$	-36.2402	-36.981	0.641
	$-1 + 0.005$	0.005	0.02525	0.02025
50	$1 + 0.005$	-36.2402	-36.7524	0.5102
	$-1 + 0.005$	0.005	0.007468	0.002468
100	$1 + 0.005$	-36.2402	-36.2961	0.0459
	$-1 + 0.005$	0.005	0.005078	0.000078

Table 28

n	z	$U(x,y)$	$U_n(x,y)$	$R_n = \lvert U - U_n\rvert$
10	$-1 + 0.005i$	0.00000	0.0202361	0.0202361
	$1 + 0.005i$	2	2.01375	0.01375
25	$-1 + 0.005i$	0.00000	0.00427	0.00427213
	$1 + 0.005i$	2	2.00215	0.00215
50	$-1 + 0.005i$	0.00000	0.00155	0.00155
	$1 + 0.005i$	2	2.00052	0.00052

To obtain the presented tables, the corresponding integral equation (6.1.7) must first be solved. Using the obtained solutions for $\left\{\varphi_n(\tau_j)\right\}_{j=1}^{n}$, an interpolation polynomial for $\psi_n(t)$ is then constructed. Finally, the approximate solution $U_n(x,y)$ is calculated using formula (6.1.11).

Despite multiple iterations of the approximation process, sufficiently high accuracy order 10^{-3} is achieved already for the $n = 100$ points.

This latter statement confirms that the approximation algorithms we have developed are stable. Rounding errors do not have a significant impact on the summation process.

In practical applications of certain types of problems (e.g., cracks), it is often difficult (sometimes impossible) to represent the integration line with exact parametric equation and only graphical form of it is known. In such cases, as demonstrated by numerical experiments, the effective method is singular integral equations method, because the computational schemes rely only on the knowledge of the nodes of the line.

Additionally, during the discretization of the equation, onle the sought solution requires approximation. In such case, we directly input the data from the monitor screen and then apply a fourth-order piecewise spline polynomial interpolation.

Using the "Spline Toolbars" package in "MatLab," the corresponding contour is generated. Along with other parameters on the monitor screen, a parameter is displayed indicating the length of the curve. The indicated contour can be further explored as a parametric equation $t = t(s)$, where the parameter s represents the length of the curve.

Additionally, it can be further explored as an analogy to the parameter of an arc. n is the number of division of the contour and $h = \dfrac{s}{n}$ is the step of division. It should be noted that the contour's division by nodes $\{s_j\}_{j=1}^{j=n}$ will turn out to be even with respect to arc length with ends s_j, s_{j+1}. Such a situation does not occur when the division of the contour is done with respect to the natural parameter.

For example, in the role of the contour L we take a line formed graphically (fig.31), and $f(t) = Re(t) + Im(t)$ as the right hand side

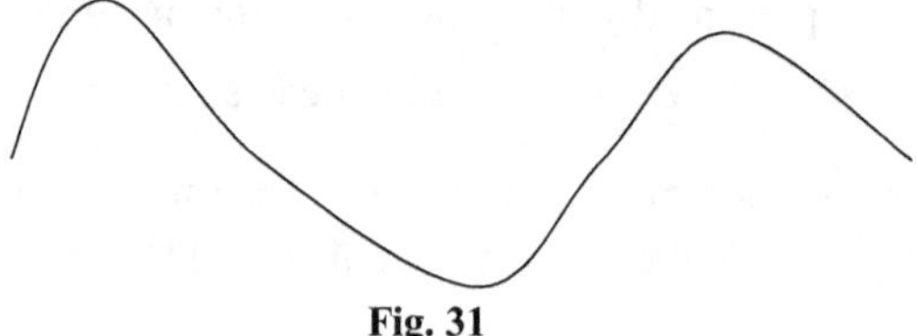

Fig. 31

and calculate the approximate solution to the Dirichlet problem for various numbers of division. The results are given in the following tables:

Table 29 $n = 10$

z	$U(x,y)$	$U_n(x,y)$	$R_n = U(x,y) - U_n(x,y)$
$005i$	$5.84648 \cdot 10^{-15}$	-0.00359624	0.00359624
$2\pi + 005i$	$1.$	0.866977	0.133003

Table 30 $n = 50$

z	$U(x,y)$	$U_n(x,y)$	$R_n = U(x,y) - U_n(x,y)$
$005i$	$5.84648 \cdot 10^{-15}$	-0.00469624	0.0046966
$2\pi + 005i$	$1.$	0.93816	0.061804

Table 31 $n = 100$

z	$U(x,y)$	$U_n(x,y)$	$R_n = U(x,y) - U_n(x,y)$
$005i$	$5.84648 \cdot 10^{-15}$	-0.00959009	0.009004
$2\pi + 005i$	$1.$	0.952251	0.0477493

It is to be noted that when dealing with regions where the contour is represented solely graphically, it is essential to estimate the increase of solution, which arises due to approximate replacing the contours. Analogous questions and generalizations in this context were first addressed by Lavrentiev, and later by other authors [40, 70].

In our case, the boundary of the domain is not given exactly, but graphically (for example, with coordinates of contour points). When solving such problems approximately, the effect of approximation of singular integrals is evident because, in such schemes, only the knowledge of the node points is needed.

§6.2 On the numerical solution of basic boundary problems on the plane in the case of linear cuts

Suppose the domain D, filled with an elastic body, represents the entire plane, which is cut along n segments of the axis Ox $L_k = a_k b_k$ $(k = 1,2,\dots,n)$. Let the set of these cuts be denoted by L.

Assume that the stress is bounded at infinity; then the functions $\Phi(z)$ and $\psi(z)$ are holomorphic in the domain D, including the point at infinity, and for large z the following relation holds [78]

$$\Phi(z) = \Gamma - \frac{X + iY}{2\pi(1 + \aleph)} \cdot \frac{1}{z} + O\left(\frac{1}{z^2}\right)$$
$$\Psi(z) = \Gamma' + \frac{\aleph(X - iY)}{2\pi(1 + \aleph)} \cdot \frac{1}{z} + O\left(\frac{1}{z^2}\right)'$$

$$(6.2.1)$$

where (X, Y) denotes the principal vector of the external load applied to the set of cuts L

$$\Gamma = B + iC, \Gamma' = B' + iC' \tag{6.2.2}$$

are constants defined by formulas [78].

$$B = \frac{1}{4}(N_1 + N_2), \qquad C = \frac{2\gamma\varepsilon_\infty}{1 + \aleph}, \qquad \Gamma' = -\frac{1}{2}(N_1 - N_2)e^{-2i\alpha}, \tag{6.2.3}$$

where N_1, N_2 are values of the principal stress at infinity, α is an angle between Ox axis and the direction of the principal stress N_1, and ε_∞ is displacement at infinity.

Let us use our notations and introduce the function

$$\Omega(z) = \overline{\Phi(z)} + z\overline{\Phi'(z)} + \overline{\Psi(z)}, \tag{6.2.4}$$

then we will have

$$Y_y - iX_y = \Phi(z) + \Omega(\bar{z}) + (z - \bar{z})\overline{\Phi'(z)} \tag{6.2.5}$$

$$2\mu(u + iv) = \aleph\varphi(z) - \omega(\bar{z}) - (z - \bar{z})\overline{\Phi(z)} + const, \tag{6.2.6}$$

where

$$\omega(z) = \int \Omega(z)dz = z\overline{\Phi(z)} + \psi(z) + const. \tag{6.2.7}$$

Hereafter, we will assume that $\Phi(z)$ and $\Omega(z)$ are piecewise holomorphic functions. Specifically, in the vicinity of any a_k, b_k endpoints, the following relationship holds:

$$|\Phi(z)| < \frac{A}{|z - c|^\alpha}, \qquad |\Omega(z)| < \frac{A}{|z - c|^\alpha}, \tag{6.2.8}$$

where A, α are positive constants, at this, $\alpha < 1$, and c denotes the corresponding endpoint. Furthermore, for each t which belongs L but does not coincide with the endpoints:

$$\lim_{z \to t} y\Phi'(z) = 0, \qquad (z = x + iy)$$

1. The first basic problem

Let us now proceed to the numerical solution of the first fundamental problem. That is, suppose we are given the values $Y_y^+, X_y^+, Y_y^-, X_y^-$ on L; the (+) and (-) signs denote, as previously mentioned, the boundary values along the crack, respectively, from above and below. We will assume that $Re\ \Gamma = B$ and $\Gamma' = B' + iC'$ are constants given beforehand. Without loss of generality, we can assume that $C = 0$ i.e. [78] $\Gamma = \bar{\Gamma} = B$.

Let's assume that the functions $p(t),\ \ q(t)$ of a class H − are given. Next, let us denote:

$$X(z) = \prod_{k=1}^{n}(z - a_k)^{\frac{1}{2}}(z - b_k)^{\frac{1}{2}} \tag{6.2.9}$$

Under $X(t)$ we will understand the limit value of the function $X(z)$ from the left of L. Then, the solution to the first basic problem is expressed as follows:

$$\Phi(z) = \Phi_0(z) + \frac{P_n(z)}{X(z)} - \frac{1}{2}\bar{\Gamma}', \quad \Omega(z) = \Omega_0(z) + \frac{P_n(z)}{X(z)} + \frac{1}{2}\Gamma'^o, \tag{6.2.10}$$

where

$$\Phi_0(z) = \frac{1}{2\pi i X(z)} \int_L \frac{X(t)p(t)dt}{t - z} + \frac{1}{2\pi i} \int_L \frac{q(t)dt}{t - z}, \tag{6.2.11}$$

$$\Omega_0(z) = \frac{1}{2\pi i X(z)} \int_L \frac{X(t)p(t)dt}{t - z} - \frac{1}{2\pi i} \int_L \frac{q(t)dt}{t - z}, \tag{6.2.12}$$

where $P_n(z)$ denotes a polynomial with order less or equal n.

$$P_n(z) = C_0 z^n + C_1 z^{n-1} + \cdots + C_n.$$

We still need to define the polynomial $P_n(t)$. We assume that under $X(z)$ we mean a branch, which for large $|z|$ takes the form

$$X(z) = +z^n + a_{n-1}z^{n-1} + \cdots \tag{6.2.13}$$

Coefficient C_0 is defined as [78]

$$C_0 = \Gamma + \frac{1}{2}\bar{\Gamma}'.$$

For determining the remaining coefficients, a system of linear equations is obtained

$$2(\aleph + 1)\int_{L_k}\frac{P_n(t)dt}{X(t)} + \aleph\int_{L_k}[\Phi_0^+(t) - \Phi_0^-(t)]dt +$$

$$+ \int_{L_k}[\Omega_0^+(t) - \Omega_0^-(t)]dt = 0 \#(6.2.14)$$

$$(k = 1, 2, \ldots, n).$$

It is proven that this system has a unique solution. As we can see, the main part of the solution formula (6.2.9) consists of Cauchy-type integrals, whose computation is challenging when the functions $p(t)$ and $q(t)$ are complex. Therefore, it is necessary to compute them approximately. In such cases, we can use the quadrature formulas developed in Chapter III, which, as shown in that chapter, compute such integrals with sufficiently high accuracy.

Thus, we can consider that the solution of the first basic problem can be approximately obtained with high-order accuracy.

2. The second basic problem

Let's assume that the displacement values are given on the upper $u^+(t)$ and lower $u^-(t)$ sides of L; moreover, if $u(a_k), v(a_k)$ and $u(b_k), v(b_k)$ denote the given displacements at the points a_k and b_k, respectively, then:

$$\begin{cases} u^+(a_k) = u^-(a_k) = u(a_k), & v^+(a_k) = v^-(a_k) = v(a_k), \\ u^+(b_k) = u^-(b_k) = u(b_k), & v^+(b_k) = v^-(b_k) = v(b_k) \end{cases} \quad (6.2.15)$$

Moreover, let's assume that Γ and Γ' are given (in this case we do not consider that $C = 0$). Additionally, we have the main vector (X, Y) associated with L. Similarly to the previous case, the general solution to the second basic problem will be:

$$\aleph\Phi(z) + \Omega(z) = \frac{1}{\pi i}\int_L \frac{g(t)dt}{t-z} + \overline{\Gamma'} + \aleph\Gamma + \overline{\Gamma}, \tag{6.2.16}$$

$$\aleph\Phi(z) - \Omega(z) = \frac{1}{\pi i X(z)}\int_L \frac{X(t)f(t)dt}{t-z} + \frac{2P_n(z)}{X(z)}, \tag{6.2.17}$$

where $X(z)$ is a function defined by formula (6.2.9), and $X(t)$ is its value from the left of L, $f(t), g(t)$ are functions of class H given on L:

$$f(t) = \mu\left[\left(u^{+\prime} + u^{-\prime}\right) + i\left(v^{+\prime} + v^{-\prime}\right)\right]$$

$$g(t) = \mu\left[\left(u^{+\prime} - u^{-\prime}\right) + i\left(v^{+\prime} - v^{-\prime}\right)\right].$$

The previous formulas define the sought after functions $\Phi(z)$ and $\psi(z)$ with polynomial approximation accuracy $P_n(z) = C_0 z^n + C_1 z^{n-1} + \cdots \ldots + C_n$. The coefficients C_0 and C_1 are determined directly from the solution, while the remaining ones are determined from the system.

$$\int_{b_k}^{a_{k+1}} [\aleph\Phi(t) - \Omega(t)]dt = 2\mu\{u(a_{k+1}) - u(b_k) + i[v(a_{k+i}) - v(b_k)]\}$$

$$(k = 1,2, \ldots, n-1).$$

Here, as in the previous case, the main part of the solution is represented by the Cauchy-type integrals, which we discussed in Chapter III. The schemes for their approximate evaluation have been presented multiple times, and we will not stop on them further here.

3. The mixed problem.

The mentioned problem was considered by D. Sherman. In this problem, the external stresses are given, which are applied, for example, to the upper side of the crack and the displacement is given on the lower side. The boundary condition on L is represented as follows:

$$\left[\Phi(t) + \frac{i}{\sqrt{\aleph}}\Omega(t)\right]^{+} - i\sqrt{\aleph}\left[\Phi(t) + \frac{i}{\sqrt{\aleph}}\Omega(t)\right]^{-} = 2f_1(t) \tag{6.2.18}$$

$$\left[\Phi(t) - \frac{i}{\sqrt{\aleph}}\Omega(t)\right]^{+} + i\sqrt{\aleph}\left[\Phi(t) - \frac{i}{\sqrt{\aleph}}\Omega(t)\right]^{-} = 2f_2(t), \tag{6.2.19}$$

where $f_1(t)$ and $f_2(t)$ are functions of class H given on L. The solution of this problem is given by formulas [78]

$$\Phi(z) + \frac{i}{\sqrt{\aleph}}\Omega(z), \qquad \Phi(z) = \frac{i}{\sqrt{\aleph}}\Omega(z)$$

$$\Phi(z) + \frac{i}{\sqrt{\aleph}}\Omega(z) = \frac{X_1(z)}{2\pi i}\int_L \frac{f_1(t)dt}{X_1^+(t)(t-z)} + X_1(z)P_n^{(1)}(z)$$

$$\Phi(z) + \frac{i}{\sqrt{\aleph}}\Omega(z) = \frac{X_2(z)}{2\pi i}\int_L \frac{f_2(t)dt}{X_2^+(t)(t-z)} + X_2(z)P_n^{(2)}(z),$$

where

$$X_1(z) = \prod_{k=1}^{n}(z - a_k)^{-\gamma_1}(z - b_k)^{-\gamma_1^{-1}},$$

$$X_2(z) = \prod_{k=1}^{n}(z - a_k)^{-\gamma_2}(z - b_k)^{-\gamma_2^{-1}}$$

At this

$$\gamma_1 = \frac{\ln(i\sqrt{\aleph})}{2\pi i} = \frac{1}{4} + \frac{\ln\aleph}{4\pi i}, \quad \gamma_2 = \frac{\ln(-i\sqrt{\aleph})}{2\pi i} = \frac{3}{4} + \frac{\ln\aleph}{4\pi i}.$$

Here, we can clearly determine $2n + 2$ coefficients of polynomials $p_n^{(1)}$ and $p_n^{(2)}$ and, accordingly, explicitly write out the functions $\Phi(z)$ and $\Omega(z)$.

§6.3 On the Numerical Solution of the Torsion Problem for a Homogeneous Prismatic Beam with cracks

According to [62], let's consider a circle with a radius r in the complex plane, with its center at the origin of the coordinates.

Let the boundary of the circle be denoted by Γ, and the diameter that coincides with the imaginary axis be denoted by Γ_2.

Let Γ_1 and Γ_1^* be the right and left halves of the circle Γ, respectively, and let L be the set of cracks L_k located on the real axis, with endpoints a_k and b_k.

$$L = \bigcup_{k=1}^{m} L_k, \qquad L_k = a_k b_k,$$

$$0 < a_1 < b_1 < a_2 < \cdots < a_k < b_k < r,$$

Let L^* denote the mirror image of the cracks L with respect to the imaginary axis.

Let the right half of the circle with cracks L be denoted by S_-, and the left half with cracks L^* by S_+; $S = S_+ \cup \Gamma_2 \cup S_-$ (fig. 32).

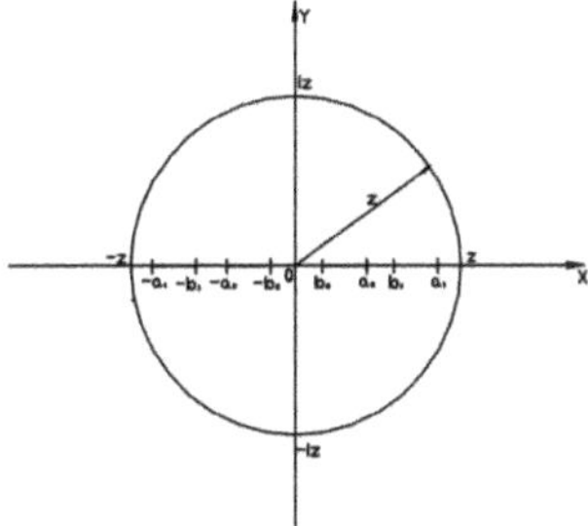

Fig. 32

Let us consider the problem [62]:

Let's find such piecewise holomorphic function $F(z) = U + IV$ in the domain S_-, which is almost bounded at corner points $t = \pm ir$, has first-order discontinuities on L and satisfies the boundary conditions

$$F(t) + \overline{F(t)} = f(t), \quad t \in \Gamma_1,$$
$$F(t) + \overline{F(t)} = \delta(t), \quad t \in \Gamma_2, \tag{6.3.1}$$
$$\left[F(t) + \overline{F(t)}\right]^{\pm} = g^{\pm}(t) + c(t), \quad t \in L, \ c(t) = c_k, t \in L_k,$$

where $f(t), \delta(t), g^{\pm}(t)$ are the given real functions belonging to class H, and c_k are real constants which are determined in the process of solving the problem.

The solution of this problem is given by formulas

$$\overline{F_1(z)} = F(z) - \overline{F(z)}$$
$$F_2(z) = F(z) + \overline{F(z)}, \qquad\qquad (6.3.2)$$
$$\overline{F(z)} = \overline{F(\bar{z})}.$$

Assume that we consider the solution to the torsion problem of a prismatic homogeneous body, where the cross-section has the shape of S_-. As it is known, ([78] § 132) this problem is reduced to the above discussed problem, i.e. to finding a piecewise holomorphic function $F(z)$ (see [78]), which is connected with the torsion function $F_*(z)$ by relation $F(z) = -iF(z)$, with the following boundary conditions

$$F(t) + \overline{F(t)} = t\bar{t}, \qquad t \in \Gamma_1, \Gamma_2,$$

$$\left[F(t) + \overline{F(t)}\right]^{\pm} = t^2 + c(t), \qquad t \in L.$$

In this concrete case the function $F(z)$ has the form

$$F(z) = -\frac{z^2}{2} + \frac{X(z)}{2\pi i} \int_L \frac{2t^2 + c(t)}{X^+(t)} Q_2(t,z)dt +$$

$$+ \frac{X(z)}{2\pi i} \int_{\Gamma_1} \frac{t\bar{t} + t^2}{X(t)} Q_2(t,z)dt, \qquad\qquad (6.3.3)$$

where $X(z)$ is a canonical function

$$X(z) = \sqrt{\prod_{k=1}^{m} (z^2 - a_k^2)(z^2 - b_k^2)\left(z^2 - \frac{r^4}{a_k^2}\right)\left(z^2 - \frac{r^4}{b_k^2}\right)},$$

and

$$Q(t,z) = \frac{1}{t-z} + \frac{1}{t+z} + \frac{z}{tz - z^2} + \frac{z}{tz + z^2}.$$

Now, let us consider a more complex case [63], where the cracks are symmetrically extended along the imiginary axis, analogously to the real axis. More precisely: a cylindrical prismatic beam with longitudinal cracks is considered. Its normal longitudinal section represents a circle with cracks along the perpendicular diameters.

Let these diameters coinside with the coordinate axes of the plane.

Let the circle with cracks $|x| < r$ be denoted by S, the circumference by Γ, and the set of cracks, which are extended correspondingly on the real and imaginary axes by L and l (see Figure 33).

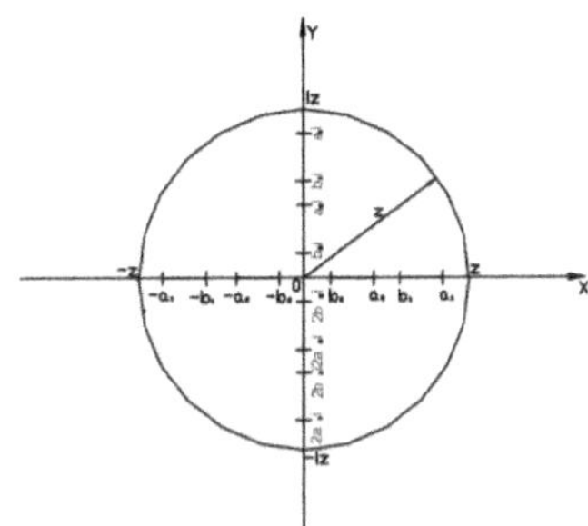

Fig. 33

$$L = \bigcup_{j=1}^{p} L_{\pm j}; \quad l = \bigcup_{k=1}^{m} l_{\pm k}; \quad \Lambda = L \bigcup l;$$

The cracks L_j and L_{-j}- are located on segments $[a_j, b_j]$ and $[-b_j, -a_j]$, and l_k and l_{-k}, correspondingly on segments $[i\alpha_k, i\beta_k]$ and $[-i\beta_k, -i\alpha_k]$.

The conditions of the Dirichlret modified problem will have the form:

$$F(t) + \overline{F(t)} = f(t), \quad t \in \Gamma$$

$$\left[F(t) + \overline{F(t)}\right]^{\pm} = \begin{cases} A^{\pm}(t) + A_j, & t \in L_j, \\ B^{\pm}(t) + A_{-j}, & j = 1,2,\cdots n. \end{cases}$$

$$\left[F(t) + \overline{F(t)}\right]^{\pm} = \begin{cases} C^{\pm}(t) + C_k & t \in L_k \\ D^{\pm}(t) + C_{-k} & t \in L_{-k}, k = 1,2,\cdots, m, \end{cases}$$

where $f(t), A^{\pm}(t), B^{\pm}(t), C^{\pm}(t), D^{\pm}(t)$ are given real functions of class H.

$$A_{\pm j}, j = 1,2,\cdots p; \quad C_{\pm k}, k = 1,2,\cdots, m$$

are arbitrary real constants determined in solution process.

In the case of torsion problem, for the following values of the given functions

$$f(t) = r^2; \quad A^+(t) = A^-(t) = B^+(t) = B^-(t) = t^2$$

$$C^+(t) = C^-(t) = D^+(t) = D^-(t) = -t^2$$

the corresponding function $F(z) = -iF_0(z)$ will have form:

$$F(z) = \frac{X(z)Q(z)}{r^{2p}r^{2m}}\left\{\frac{r^2}{4\pi i}\int_\Gamma \frac{r^{2p}r^{2m}}{X(\tau)Q(\tau)}\left[\frac{r}{r^2 - z^2} - \frac{rz^2}{t^2z^2 - r^2}\right]d\tau\right\} +$$

$$+\frac{X(z)Q(z)}{r^{2p}r^{2m}}\left\{\frac{1}{8\pi i}\int_\Lambda \frac{t^{2p}t^{2m}[g_1(t) + D_1(t)]}{X^+(t)Q^+(t)}\left[\frac{1}{t^2 - z^2} - \frac{tz^2}{t^2z^2 - r^4}\right]d\tau\right\} + iK_0$$

$$(6.3.4)$$

where K_0 is an arbitrary function.

$$g_1(t) = 4t^2, D_1(t) = 4A_j, A_j = A_{-j}, t \in L_j$$

$$j = 1,2,\cdots p$$

$$g_1(t) = -4t^2, D_1(t) = 4C_k, C_k = C_{-k}, t \in l_k$$

$$k = 1,2,\cdots m;$$

After that, the stress function can be determined and, accordingly, the stresses of the prismatic body under the action of the cracks can be derived, as outlined in [78]. Similarly, it is possible to determine the intensity coefficients at the crack tips. Further detailed cases are discussed in the references [39, 110].

As we see, the determination of torsion functions is quite complicated. In practice, the exact calculation is difficult, so it is important to compute such integrals with a high order accuracy, as in this case, the use of high-order approximations helps to solve torsion problem. For formulas (6.3.3) and (6.3.4), we can compute the approximations using quadrature formulas previously developed for approximating Cauchy-type integrals with weight. Additionally, the conditions of solvability must be satisfied [62,63].

Appendix

Discrete Singularities Method

```
n = 50
Do[τ_σ = -1 + 2 (σ - 1) / n, {σ, 1, n}]
τ_{n+1} = 1

50

1

Do[t_j = (τ_j + τ_{j+1}) / 2, {j, 1, n}]

Do[p_σ = (1 / π) ∫_{τ_σ}^{τ_{σ+1}} √((1 + t) / (1 - t)) ((τ_{σ+1} - t) / (τ_{σ+1} - τ_σ)) dt, {σ, 1, n}]

Do[p1_σ = (1 / π) ∫_{τ_σ}^{τ_{σ+1}} √((1 + t) / (1 - t)) ((τ_σ - t) / (τ_σ - τ_{σ+1})) dt, {σ, 1, n}]

φ[t_, τ_] := 1 / (t - τ)

Table[N[p_1 φ[τ_1, t_j] + Σ_{σ=1}^{n-2} (p_{σ+1} + p1_σ) φ[τ_{σ+1}, t_j] + p1_{n-1} φ[τ_n, t_j]], {j, 1, n}]
```

{0.90243, 0.906759, 0.904852, 0.902803, 0.900654, 0.898405, 0.896052, 0.893586,
0.891001, 0.888287, 0.885434, 0.882431, 0.879267, 0.875927, 0.872398, 0.868661,
0.8647, 0.860492, 0.856013, 0.851238, 0.846134, 0.840668, 0.8348, 0.828482, 0.821661,
0.814276, 0.806252, 0.797503, 0.787925, 0.777397, 0.765767, 0.752855, 0.738435,
0.722225, 0.703872, 0.682918, 0.658768, 0.63063, 0.597423, 0.55764, 0.509104, 0.448557,
0.370887, 0.267574, 0.123256, -0.0929981, -0.454816, -1.1978, -3.98369, -5.10961}

Approximation of Cauchy Integral

```mathematica
n = 5
L : Do[τ_σ = -1 + 2 (σ - 1) / n, {σ, 1, n}]
τ_{n+1} = 1

5

L : Null

1

x = 1
a = -1
b = 1
z = 1 + 0.005 I
t_θ = 1
v = n

1

-1

1

1 + 0.005 i

1

5

φ[t_] := Re[t] + Im[ t]
L :   l[v_, 0 _, t_] := (t - τ_{v+1}) / (τ_v - τ_{v+1})
      l[v_, 1 _, t_] := (t - τ_v) / (τ_{v+1} - τ_v)
      L[v_, φ_, t_] := l[v, 0, t] φ[τ_v] + l[v, 1, t] φ[τ_{v+1}]
χ[z_, a_, b_] := √((z - a) (b - z))

F[φ_, z_, a_, b_] := (1 / (π*I χ[z, a, b])) ∫_a^b φ[t] χ[t, a, b] / ((t - z)) dt

F[φ, z, a, b]

-3.60173 + 3.46915 i

L :
  Do[p[σ_, k_, x_, z_] := (1 / (π*I)) ∫_{τ_σ}^{τ_{σ+1}} (t - x) χ[t, a, b] l[σ, k, t] / ((t - z)) dt,

  {σ, 1, n}, {k, 0, 1}]

N[(1 / (π*I χ[z, a, b])) ∫_a^b (χ[t, a, b] / (t - z)) dt] L[v, φ, t_θ] +

  (1 / χ[z, a, b]) (Σ_{σ=1}^{n-2} Σ_{k=0}^{1} p[σ, k, x, z] (φ[τ_{σ+k}] - φ[t_θ]) / (τ_{σ+k} - z) +

    p[v, 0, 1, z] (φ[τ_v] - L[v, φ, t_θ]) / (τ_v - z) + p[v, 1, 1, z] (φ[τ_{v+1}] - L[v, φ, t_θ]) / (τ_{v+1} - z) +

    p[v - 1, 1, 1, z] (φ[τ_v] - L[v, φ, t_θ]) / (τ_v - z) + p[v - 1, 0, 1, z] (φ[τ_{v-1}] - φ[t_θ]) / (τ_{v-1} - z))

L : Null

-3.8148 + 3.66829 i
```

Airfoil Problem

n = 5

5

$$u2[x\_, y\_] := \sum_{k=-\infty}^{\infty} (1/2\,\pi)\, \frac{y-x}{(y-x)\,\hat{}\,2 + (k*1)\,\hat{}\,2}$$

$$u1[x\_, y\_] := (\pi/1) * ((y-x) * u2[x, y] - 1/\pi)/(y-x)$$

$$s1[v\_, j\_, n\_, m\_] := \sum_{\sigma=1}^{v-1}\left(\sum_{k=1}^{m-1} pp[\sigma, k]/(\lambda_{v,j} - \lambda_{\sigma,k})\right) + \sum_{\sigma=v+1}^{n}\left(\sum_{k=1}^{m-1} pp[\sigma, k]/(\lambda_{v,j} - \lambda_{\sigma,k})\right);$$

$$s2[v\_, j\_, m\_] := \sum_{k=1}^{j-1} pp[v, k]/(\lambda_{v,j} - \lambda_{v,k}) + \sum_{k=j+1}^{m-1} pp[v, k]/(\lambda_{v,j} - \lambda_{v,k});$$

$$s3[v\_, j\_, m\_] := pp[v, j]\left(\sum_{k=1}^{j-1} dd[v, j, k, m] + \sum_{k=j+1}^{m} dd[v, j, k, m]\right);$$

$$s4[v\_, j\_] := pp[v, j] \cdot u1[\lambda_{v,j}, \lambda_{v,j}];$$

$$s5[v\_, j\_, m\_] := \sum_{k=1}^{j-1} (pp[v, k]/(\lambda_{v,j} - \lambda_{v,k}) - pp[v, k] \cdot u1[\lambda_{v,j}, \lambda_{v,k}])\,\varphi[\lambda_{v,k}] +$$

$$\sum_{k=j+1}^{m-1} (pp[v, k]/(\lambda_{v,j} - \lambda_{v,k}) - pp[v, k] \cdot u1[\lambda_{v,j}, \lambda_{v,k}])\,\varphi[\lambda_{v,k}];$$

$$s6[v\_, j\_, m\_] :=$$

$$\left(\sum_{k=1}^{j-1} pp[v, j] \cdot dd[v, j, k, m] \cdot \varphi[\lambda_{v,k}] + \sum_{k=j+1}^{m} pp[v, j] \cdot dd[v, j, k, m] \cdot \varphi[\lambda_{v,k}]\right)$$

$$s7[v\_, j\_, n\_, m\_] := \sum_{\sigma=1}^{v-1}\sum_{k=1}^{m-1} (pp[\sigma, k]/(\lambda_{v,j} - \lambda_{\sigma,k}) - pp[\sigma, k] \cdot u1[\lambda_{v,j}, \lambda_{\sigma,k}])\,\varphi[\lambda_{\sigma,k}] +$$

$$\sum_{\sigma=v+1}^{n}\sum_{k=1}^{m-1} (pp[\sigma, k]/(\lambda_{v,j} - \lambda_{\sigma,k}) - pp[\sigma, k] \cdot u1[\lambda_{v,j}, \lambda_{\sigma,k}])\,\varphi[\lambda_{\sigma,k}];$$

Do[τ_σ = -1 + 2 (σ - 1)/n, {σ, 1, n}]

Table[τ_σ, {σ, 1, n}]

$$\left\{-1, -\frac{3}{5}, -\frac{1}{5}, \frac{1}{5}, \frac{3}{5}\right\}$$

m = 4

4

```mathematica
Do[x_k = (k - 1) / (m - 1), {k, 1, m}]
Table[x_k, {k, 1, m}]
```

$$\left\{0, \frac{1}{3}, \frac{2}{3}, 1\right\}$$

```mathematica
h = 2 / n
```

$$\frac{2}{5}$$

```mathematica
Do[λ_{σ,k} = τ_σ + h x_k, {σ, 1, n}, {k, 1, m}]
Table[N[λ_{σ,k}], {σ, 1, n}, {k, 1, m}]
```

{{-1., -0.866667, -0.733333, -0.6}, {-0.6, -0.466667, -0.333333, -0.2},
 {-0.2, -0.0666667, 0.0666667, 0.2}, {0.2, 0.333333, 0.466667, 0.6}, {0.6, 0.733333, 0.866667, 1.}}

```mathematica
ω[t_, v_, m_] := Product[t - λ_{v,k}, {k, 1, m}]

τ_{n+1} = 1
Do[p[σ_, k_] := (1 / (π D[ω[t, σ, m], t] /. t → λ_{σ,k}))
    Integrate[ω[t, σ, m] / (√((1 + t) / (1 - t)) (t - λ_{σ,k})), {t, τ_σ, τ_{σ+1}}], {σ, 1, n}, {k, 1, m}]
```

1

```mathematica
Table[N[p[σ, k]], {σ, 1, n}, {k, 1, m}]
```

{{0.18328, 0.239118, 0.0874512, 0.0399664},
 {0.029727, 0.0844064, 0.0630042, 0.0208318}, {0.0185239, 0.0537195, 0.0421604, 0.0137846},
 {0.0123011, 0.0357865, 0.0267983, 0.00862186}, {0.00719667, 0.0211938, 0.00989287, 0.00223595}}

```mathematica
pp[1, 1] = N[p[1, 1]]
pp[n, m] = N[p[n, m]]
```

0.18328

0.00223595

```mathematica
pp[n - 1, m] = N[p[n, 1]]
```

0.00719667

```mathematica
Do[pp[σ, m - 1] = pp[σ, 0], {σ, 1, n - 1}]
Do[pp[σ, k] = p[σ, k], {σ, 1, n}, {k, 1, m - 1}]
Do[pp[σ, 1] = p[σ - 1, m] + p[σ, 1], {σ, 2, n - 1}]
Do[pp[σ - 1, m] = pp[σ, 1], {σ, 2, n}]
Do[pp[n, k] = N[p[n, k]], {k, 1, m}]


Table[N[pp[σ, k]], {σ, 1, n}, {k, 1, m}]
```

{{0.18328, 0.239118, 0.0874512, 0.0696934},
 {0.0696934, 0.0844064, 0.0630042, 0.0393558}, {0.0393558, 0.0537195, 0.0421604, 0.0260857},
 {0.0260857, 0.0357865, 0.0267983, 0.00719667}, {0.00719667, 0.0211938, 0.00989287, 0.00223595}}

```
dd[v_, j_, k_, m_] :=
```

$$\left(\texttt{Delete}\left[\prod_{i=1}^{m}(\lambda_{v,j}-h_{v,i}),\ \{\{j\},\ \{k\}\}\right]\,/.\,h\to\lambda\right)\Bigg/\left(\texttt{Delete}\left[\prod_{i=1}^{m}(\lambda_{v,k}-h_{v,i}),\ \{\{k\}\}\right]\,/.\,h\to\lambda\right)$$

```
lag[f_, x_, m_] :=
```

$$\sum_{k=1}^{m}(\texttt{Product}[x-h_i,\ \{i,\ 1,\ m\}]\ f[h_k])\,/\,((x-h_k)\ (\texttt{D}[\texttt{Product}[x-h_i,\ \{i,\ 1,\ m\}],\ x]\,/.\,x\to h_k))$$

```
Do[Print[N[(1 + s1[v, j, n, m] + s2[v, j, m] - s3[v, j, m] + s4[v, j]) φ[λv,j] -
    s5[v, j, m] + s6[v, j, m] - s7[v, j, n, m]] == 2], {v, 1, n - 1}, {j, 1, m - 1}]

Do[Print[N[(1 + s1[n, j, n, m] + s2[n, j, m] - s3[n, j, m] + s4[n, j]) φ[λn,j] - s5[n, j, m] + s6[n, j, m] -
    s7[n, j, n, m]] == 2], {j, 1, m}]

NSolve[{-4.363240434116095` φ[-1.`] + 5.848514927765674` φ[-0.8666666666666667`] -
    1.7553327166614825` φ[-0.7333333333333333`] + 0.6185748253330091` φ[-0.6`] +
    0.14543324222344625` φ[-0.4666666666666667`] + 0.08785806705809689` φ[-0.3333333333333333`] +
    0.04679721959097781` φ[-0.2`] + 0.05651635455050569` φ[-0.06666666666666667`] +
    0.04029802265059015` φ[0.06666666666666667`] + 0.023086586122036883` φ[0.2`] +
    0.029724680940943337` φ[0.3333333333333333`] + 0.021090164253844076` φ[0.4666666666666667`] +
    0.005402331320441883` φ[0.6`] + 0.015246665320053887` φ[0.7333333333333333`] +
    0.006842942900556887` φ[0.8666666666666667`] == 2,
  -2.025022465839446` φ[-1.`] - 0.09651832870748861` φ[-0.8666666666666667`] +
    2.4318779231777192` φ[-0.7333333333333333`] - 0.0481395305896955` φ[-0.6`] +
    0.20210942376709495` φ[-0.4666666666666667`] + 0.11429467255199971` φ[-0.3333333333333333`] +
    0.05827155152861302` φ[-0.2`] + 0.06813379786079456` φ[-0.06666666666666667`] +
    0.04735146243826407` φ[0.06666666666666667`] + 0.02655811108028354` φ[0.2`] +
    0.03358602542518698` φ[0.3333333333333333`] + 0.023466508027629604` φ[0.4666666666666667`] +
    0.005932113823920398` φ[0.6`] + 0.016552189385780725` φ[0.7333333333333333`] +
    0.007356234942021908` φ[0.8666666666666667`] == 2,
  -0.6227869609200843` φ[-1.`] - 2.4968142714514636` φ[-0.8666666666666667`] +
    2.4816647999466226` φ[-0.7333333333333333`] + 0.7339743349926515` φ[-0.6`] +
    0.31138200069750954` φ[-0.4666666666666667`] + 0.15629034285456994` φ[-0.3333333333333333`] +
    0.07451327550077451` φ[-0.2`] + 0.08335636453702555` φ[-0.06666666666666667`] +
    0.05609918736573235` φ[0.06666666666666667`] + 0.030692549876943573` φ[0.2`] +
    0.03804716908867127` φ[0.3333333333333333`] + 0.026149826034791126` φ[0.4666666666666667`] +
    0.006520111032777229` φ[0.6`] + 0.01798265632502894` φ[0.7333333333333333`] +
    0.007913367684342516` φ[0.8666666666666667`] == 2,
  -0.4946436230497585` φ[-1.`] - 0.933034492530502` φ[-0.8666666666666667`] -
    0.6651117030997664` φ[-0.7333333333333333`] + 0.7918669359117113` φ[-0.6`] +
    2.1995144505655406` φ[-0.4666666666666667`] - 0.5466304135866762` φ[-0.3333333333333333`] +
    0.27465744497537947` φ[-0.2`] + 0.10505448051058627` φ[-0.06666666666666667`] +
    0.06767501957728446` φ[0.06666666666666667`] + 0.03588528401966831` φ[0.2`] +
    0.043441101058788574` φ[0.3333333333333333`] + 0.029303690611939298` φ[0.4666666666666667`] +
```

```
   0.00719666682350275` φ[0.6`] + 0.0196023217940717` φ[0.7333333333333333`] +
   0.008536587498919683` φ[0.8666666666666667`] == 2,
-0.37150585147259624` φ[-1.`] - 0.6230259660145641` φ[-0.8666666666666667`] -
   0.3332693415369052` φ[-0.7333333333333333`] - 0.7350658845622866` φ[-0.6`] +
   1.4992422245013644` φ[-0.4666666666666667`] + 1.1088360676326385` φ[-0.3333333333333333`] +
   0.04524870654300564` φ[-0.2`] + 0.13994884771354937` φ[-0.06666666666666667`] +
   0.08434919473664171` φ[0.06666666666666667`] + 0.042844852701737504` φ[0.2`] +
   0.05031546074580114` φ[0.3333333333333333`] + 0.03317885543716322` φ[0.4666666666666667`] +
   0.008005617878304214` φ[0.6`] + 0.021499425946793285` φ[0.7333333333333333`] +
   0.009255149613742847` φ[0.8666666666666667`] == 2,
-0.2942592776902384` φ[-1.`] - 0.4629124771247091` φ[-0.8666666666666667`] -
   0.22032141119694215` φ[-0.7333333333333333`] - 0.18131784038560742` φ[-0.6`] -
   1.1012154294203977` φ[-0.4666666666666667`] + 2.3037062003605677` φ[-0.3333333333333333`] +
   0.45681801402711736` φ[-0.2`] + 0.2081991088363017` φ[-0.06666666666666667`] +
   0.11140763408848331` φ[0.06666666666666667`] + 0.05297971008570927` φ[0.2`] +
   0.059644149362366346` φ[0.3333333333333333`] + 0.038185076855602754` φ[0.4666666666666667`] +
   0.009013923229202758` φ[0.6`] + 0.023801484615374152` φ[0.7333333333333333`] +
   0.010109418979755235` φ[0.8666666666666667`] == 2,
-0.2402648666956958` φ[-1.`] - 0.3633070189507744` φ[-0.8666666666666667`] -
   0.16236840292646726` φ[-0.7333333333333333`] - 0.17063053989849403` φ[-0.6`] -
   0.3097197548222555` φ[-0.4666666666666667`] - 0.4659044999296667` φ[-0.3333333333333333`] +
   1.4043526962514106` φ[-0.2`] + 1.2960543451926276` φ[-0.06666666666666667`] -
   0.27807410853193415` φ[0.06666666666666667`] + 0.16795116395003445` φ[0.2`] +
   0.07335895233584815` φ[0.3333333333333333`] + 0.04505023212530538` φ[0.4666666666666667`] +
   0.010331085766465993` φ[0.6`] + 0.026703924123529843` φ[0.7333333333333333`] +
   0.011158091411109534` φ[0.8666666666666667`] == 2,
-0.19992032294081988` φ[-1.`] - 0.29451528283929485` φ[-0.8666666666666667`] -
   0.12665584259463614` φ[-0.7333333333333333`] - 0.12505698367983725` φ[-0.6`] -
   0.20213824489141063` φ[-0.4666666666666667`] - 0.22834776024807388` φ[-0.3333333333333333`] -
   0.4238600678520946` φ[-0.2`] + 1.7048772831806063` φ[-0.06666666666666667`] +
   0.7261263045607043` φ[0.06666666666666667`] + 0.03523437406390334` φ[0.2`] +
   0.09594662232455065` φ[0.3333333333333333`] + 0.055218917578975825` φ[0.4666666666666667`] +
   0.012151977142353186` φ[0.6`] + 0.03052745564128345` φ[0.7333333333333333`] +
   0.0124916092141425` φ[0.8666666666666667`] == 2,
-0.16846614548104183` φ[-1.`] - 0.24383599807504122` φ[-0.8666666666666667`] -
   0.1022643852877747` φ[-0.7333333333333333`] - 0.09720986542574599` φ[-0.6`] -
   0.1476544929733417` φ[-0.4666666666666667`] - 0.14853430043278493` φ[-0.3333333333333333`] -
   0.088744455653288` φ[-0.2`] - 0.7101463142592783` φ[-0.06666666666666667`] +
   2.219116859363129` φ[0.06666666666666667`] + 0.30576723175222315` φ[0.2`] +
   0.1408390751961005` φ[0.3333333333333333`] + 0.07204875541499521` φ[0.4666666666666667`] +
   0.014863941791825507` φ[0.6`] + 0.03584429847964844` φ[0.7333333333333333`] +
   0.0142589680938907` φ[0.8666666666666667`] == 2,
-0.14325820350246302` φ[-1.`] - 0.20489718251396977` φ[-0.8666666666666667`] -
   0.08449976108614816` φ[-0.7333333333333333`] - 0.07835827015083907` φ[-0.6`] -
   0.1145843453519247` φ[-0.4666666666666667`] - 0.10830474451471586` φ[-0.3333333333333333`] -
   0.09183012858164673` φ[-0.2`] - 0.19205222838339592` φ[-0.06666666666666667`] -
   0.30856871835303673` φ[0.06666666666666667`] + 1.6232041155588677` φ[0.2`] +
   0.8620736256075595` φ[0.3333333333333333`] - 0.1878677469836783` φ[0.4666666666666667`] +
   0.08458223197643949` φ[0.6`] + 0.043793651430553006` φ[0.7333333333333333`] +
   0.016726645796602407` φ[0.8666666666666667`] == 2,
-0.12268533184614229` φ[-1.`] - 0.174114875062987` φ[-0.8666666666666667`] -
```

```
  0.07099538252943219` φ[-0.7333333333333333`] - 0.06474239727314816` φ[-0.6`] -
  0.09234138459204515` φ[-0.4666666666666667`] - 0.08400553770660583` φ[-0.3333333333333333`] -
  0.06690849360705918` φ[-0.2`] - 0.12457094195619295` φ[-0.06666666666666667`] -
  0.15027932477257513` φ[0.06666666666666667`] - 0.28019000482323264` φ[0.2`] +
  1.828336295824159` φ[0.3333333333333333`] + 0.4745114088295319` φ[0.4666666666666667`] -
  0.016368614158922514` φ[0.6`] + 0.05702793766306539` φ[0.7333333333333333`] +
  0.020425858417864356` φ[0.8666666666666667`] == 2,
-0.1056864991982783` φ[-1.`] - 0.1492880762119106` φ[-0.8666666666666667`] -
  0.06041698370397232` φ[-0.7333333333333333`] - 0.05446604823185417` φ[-0.6`] -
  0.07636767244608102` φ[-0.4666666666666667`] - 0.06773536507108496` φ[-0.3333333333333333`] -
  0.05190691259806615` φ[-0.2`] - 0.09075706874885829` φ[-0.06666666666666667`] -
  0.09745149702209382` φ[0.06666666666666667`] - 0.05935680161794716` φ[0.2`] -
  0.462541451798055 9` φ[0.3333333333333333`] + 2.1905217386659` φ[0.4666666666666667`] +
  0.12234373467771216` φ[0.6`] + 0.08349750776230117` φ[0.7333333333333333`] +
  0.026594243906862734` φ[0.8666666666666667`] == 2,
-0.09151686102614595` φ[-1.`] - 0.12896817936702698` φ[-0.8666666666666667`] -
  0.05194684490533022` φ[-0.7333333333333333`] - 0.04646225546509773` φ[-0.6`] -
  0.06436776539002464` φ[-0.4666666666666667`] - 0.05609533966191044` φ[-0.3333333333333333`] -
  0.04189273703271961` φ[-0.2`] - 0.07045009408501313` φ[-0.06666666666666667`] -
  0.07102380285803041` φ[0.06666666666666667`] - 0.06022499436040105` φ[0.2`] -
  0.12735201356403214` φ[0.3333333333333333`] - 0.19587476258700032` φ[0.4666666666666667`] +
  1.8634708887104343` φ[0.6`] + 0.3248680300581591` φ[0.7333333333333333`] -
  0.042020022764338735` φ[0.8666666666666667`] + 0.017991667058756877` φ[1.`] == 2,
-0.07962681803515902` φ[-1.`] - 0.11214852465313006` φ[-0.8666666666666667`] -
  0.04505061151687245` φ[-0.7333333333333333`] - 0.04008022207551011` φ[-0.6`] -
  0.05505391302591118` φ[-0.4666666666666667`] - 0.04737675477737607` φ[-0.3333333333333333`] -
  0.03474611730636553` φ[-0.2`] - 0.056921579859171394` φ[-0.06666666666666667`] -
  0.05517706950692131` φ[0.06666666666666667`] - 0.04391941694356201` φ[0.2`] -
  0.08263893912430051` φ[0.3333333333333333`] - 0.0954098622918481` φ[0.4666666666666667`] -
  0.1056050284526524 2` φ[0.6`] + 1.8121677602835544` φ[0.7333333333333333`] +
  0.2349742300196505` φ[0.8666666666666667`] - 0.02649230619312299` φ[1.`] == 2,
-0.06959571746137332` φ[-1.`] - 0.09809959294905435` φ[-0.8666666666666667`] -
  0.0393612239208182` φ[-0.7333333333333333`] - 0.034897785997963886` φ[-0.6`] -
  0.0476442255255458` φ[-0.4666666666666667`] - 0.04062363070224122` φ[-0.3333333333333333`] -
  0.029402989881970557` φ[-0.2`] - 0.0472822213768 5143` φ[-0.06666666666666667`] -
  0.0446336536367028` φ[0.06666666666666667`] - 0.03415190307651233` φ[0.2`] -
  0.06031077374001166` φ[0.3333333333333333`] - 0.06195168251947184` φ[0.4666666666666667`] -
  0.013279815682609643` φ[0.6`] - 0.22924297001859062` φ[0.7333333333333333`] +
  1.9993399395791271` φ[0.8666666666666667`] + 0.02473216920162538` φ[1.`] == 2,
-0.061093265867495174` φ[-1.`] -
  0.08627558442294134` φ[-0.8666666666666667`] - 0.03461650529158961` φ[-0.7333333333333333`] -
  0.030627621960754588` φ[-0.6`] - 0.041634446823791485` φ[-0.4666666666666667`] -
  0.035257595747418914` φ[-0.3333333333333333`] -
  0.02526929698093911` φ[-0.2`] - 0.04008343571277299` φ[-0.06666666666666667`] -
  0.037128609058598404` φ[0.06666666666666667`] - 0.027658473779664895` φ[0.2`] -
  0.04694385972761588` φ[0.3333333333333333`] - 0.045251092732617765` φ[0.4666666666666667`] -
  0.0222547304886647/9` φ[0.6`] + 0.05046281329115727 4` φ[0.7333333333333333`] -
  0.12272736983065292` φ[0.8666666666666667`] + 1.886222234290101` φ[1.`] == 2}]
```



```
{{φ[-1.] → 3.1792, φ[-0.866667] → 3.09753, φ[-0.733333] → 3.00583, φ[-0.6] → 2.90771,
  φ[-0.466667] → 2.80635, φ[-0.333333] → 2.705, φ[-0.2] → 2.60631, φ[-0.0666667] → 2.51229,
  φ[0.0666667] → 2.4246, φ[0.2] → 2.34412, φ[0.333333] → 2.27128, φ[0.466667] → 2.20614,
  φ[0.6] → 2.14843, φ[0.733333] → 2.0977, φ[0.866667] → 2.0534, φ[1.] → 2.01479}}
```

Solving Singular Integral Equation

```
n = 3

3

u1[x_, y_] := (y^2 + 6 y + 4 x y + 2 x - x^2 + 4) / (x + y + 2) ^3

u1[x_, y_] = 1

1
```

$$s1[v\_, j\_, n\_, m\_] := \sum_{\sigma=1}^{v-1}\left(\sum_{k=1}^{m-1} pp[\sigma, k] / (\lambda_{v,j} - \lambda_{\sigma,k})\right) + \sum_{\sigma=v+1}^{n}\left(\sum_{k=1}^{m-1} pp[\sigma, k] / (\lambda_{v,j} - \lambda_{\sigma,k})\right);$$

$$s2[v\_, j\_, m\_] := \sum_{k=1}^{j-1} pp[v, k] / (\lambda_{v,j} - \lambda_{v,k}) + \sum_{k=j+1}^{m-1} pp[v, k] / (\lambda_{v,j} - \lambda_{v,k});$$

$$s3[v\_, j\_, m\_] := pp[v, j]\left(\sum_{k=1}^{j-1} dd[v, j, k, m] + \sum_{k=j+1}^{m} dd[v, j, k, m]\right);$$

$$s4[v\_, j\_] := pp[v, j] \times u1[\lambda_{v,j}, \lambda_{v,j}];$$

$$s5[v\_, j\_, m\_] := \sum_{k=1}^{j-1} (pp[v, k] / (\lambda_{v,j} - \lambda_{v,k}) - pp[v, k] \times u1[\lambda_{v,j}, \lambda_{v,k}]) \, \varphi[\lambda_{v,k}] +$$
$$\sum_{k=j+1}^{m-1} (pp[v, k] / (\lambda_{v,j} - \lambda_{v,k}) - pp[v, k] \times u1[\lambda_{v,j}, \lambda_{v,k}]) \, \varphi[\lambda_{v,k}];$$

$$s6[v\_, j\_, m\_] :=$$
$$\left(\sum_{k=1}^{j-1} pp[v, j] \times dd[v, j, k, m] \times \varphi[\lambda_{v,k}] + \sum_{k=j+1}^{m} pp[v, j] \times dd[v, j, k, m] \times \varphi[\lambda_{v,k}]\right)$$

$$s7[v\_, j\_, n\_, m\_] := \sum_{\sigma=1}^{v-1}\sum_{k=1}^{m-1} (pp[\sigma, k] / (\lambda_{v,j} - \lambda_{\sigma,k}) - pp[\sigma, k] \times u1[\lambda_{v,j}, \lambda_{\sigma,k}]) \, \varphi[\lambda_{\sigma,k}] +$$
$$\sum_{\sigma=v+1}^{n}\sum_{k=1}^{m-1} (pp[\sigma, k] / (\lambda_{v,j} - \lambda_{\sigma,k}) - pp[\sigma, k] \times u1[\lambda_{v,j}, \lambda_{\sigma,k}]) \, \varphi[\lambda_{\sigma,k}];$$

```
Do[τ_σ = -1 + 2 (σ - 1) / n, {σ, 1, n}]

m = 2

2

Do[x_k = (k - 1) / (m - 1), {k, 1, m}]
Table[x_k, {k, 1, m}]

{0, 1}

h = 2 / n

2
─
3
```

```mathematica
Do[λ_{σ,k} = τ_σ + h x_k, {σ, 1, n}, {k, 1, m}]
Table[N[λ_{σ,k}], {σ, 1, n}, {k, 1, m}]
```

$\{\{-1., -0.333333\}, \{-0.333333, 0.333333\}, \{0.333333, 1.\}\}$

```mathematica
ω[t_, v_, m_] := Product[t - λ_{v,k}, {k, 1, m}]

τ_{n+1} = 1
Do[p[σ_, k_] := (1 / (π D[ω[t, σ, m], t] /. t → λ_{σ,k}))
    Integrate[ω[t, σ, m] / ( √((1 + t) / (1 - t)) (t - λ_{σ,k})), {t, τ_σ, τ_{σ+1}}], {σ, 1, n}, {k, 1, m}]
```

1

```mathematica
pp[1, 1] = N[p[1, 1]]
pp[n, m] = N[p[n, m]]
```

0.473088

0.0354013

```mathematica
Do[pp[σ, m - 1] = pp[σ, 0], {σ, 1, n - 1}]
Do[pp[σ, k] = p[σ, k], {σ, 1, n}, {k, 1, m - 1}]
Do[pp[σ, 1] = p[σ - 1, m] + p[σ, 1], {σ, 2, n - 1}]
Do[pp[σ - 1, m] = pp[σ, 1], {σ, 2, n}]
Do[pp[n, k] = N[p[n, k]], {k, 1, m}]

Table[N[pp[σ, k]], {σ, 1, n}, {k, 1, m}]
```

$\{\{0.473088, 0.339224\}, \{0.339224, 0.0563198\}, \{0.0563198, 0.0354013\}\}$

```mathematica
dd[v_, j_, k_, m_] :=
  (Delete[∏_{i=1}^{m} (λ_{v,j} - h_{v,i}), {{j}, {k}}] /. h → λ) / (Delete[∏_{i=1}^{m} (λ_{v,k} - h_{v,i}), {{k}}] /. h → λ)

Do[Print[N[(1 + s1[v, j, n, m] + s2[v, j, m] - s3[v, j, m] + s4[v, j]) φ[λ_{v,j}] -
    s5[v, j, m] + s6[v, j, m] - s7[v, j, n, m]] == 2], {v, 1, n - 1}, {j, 1, m - 1}]

Do[Print[N[(1 + s1[n, j, n, m] + s2[n, j, m] - s3[n, j, m] + s4[n, j]) φ[λ_{n,j}] - s5[n, j, m] + s6[n, j, m] -
    s7[n, j, n, m]] == 2], {j, 1, m}]
```

$0.212379\, \varphi[-1.] + 1.55769\, \varphi[-0.333333] + 0.0985596\, \varphi[0.333333] == 2$

$-0.236544\, \varphi[-1.] + 1.45554\, \varphi[-0.333333] + 0.649636\, \varphi[0.333333] == 2$

$0.118272\, \varphi[-1.] - 0.169612\, \varphi[-0.333333] + 1.83549\, \varphi[0.333333] + 0.0844797\, \varphi[1.] == 2$

$0.236544\, \varphi[-1.] + 0.0848061\, \varphi[-0.333333] - 0.0812619\, \varphi[0.333333] + 1.66395\, \varphi[1.] == 2$


```mathematica
NSolve[{0.2123792602423965` φ[-1.`] + 1.5576937918775713` φ[-0.3333333333333333`] +
    0.09855962478515816` φ[0.3333333333333333`] == 2,
  -0.23654421820322308` φ[-1.`] + 1.4555407487688103` φ[-0.3333333333333333`] +
    0.6496361463395387` φ[0.3333333333333333`] == 2,
  0.11827210910161154` φ[-1.`] - 0.16961222745358046` φ[-0.3333333333333333`] +
    1.8354931168698165` φ[0.3333333333333333`] + 0.08447967838727842` φ[1.`] == 2,
  0.23654421820322308` φ[-1.`] + 0.0848061137267902` φ[-0.3333333333333333`] -
    0.08126188437583878` φ[0.3333333333333333`] + 1.6639455570710042` φ[1.`] == 2},
  {φ[-1.`], φ[-0.3333333333333333`], φ[0.3333333333333333`], φ[1.`]}]
```

$\{\{\varphi[-1.] \to 1.07152, \varphi[-0.333333] \to 1.07007, \varphi[0.333333] \to 1.07125, \varphi[1.] \to 1.04741\}\}$

217

```
n = 11
```

11

```
u1[x_, y_] := (x + y + 4 Cos[π / 3]) / ((x + y + 4 Cos[π / 6]) ^2 + 16 (Sin[π / 3]) ^2)
```

$$s1[v\_, j\_, n\_, m\_] := \sum_{\sigma=1}^{v-1} \left(\sum_{k=1}^{m-1} pp[\sigma, k] / (\lambda_{v,j} - \lambda_{\sigma,k}) \right) + \sum_{\sigma=v+1}^{n} \left(\sum_{k=1}^{m-1} pp[\sigma, k] / (\lambda_{v,j} - \lambda_{\sigma,k}) \right);$$

$$s2[v\_, j\_, m\_] := \sum_{k=1}^{j-1} pp[v, k] / (\lambda_{v,j} - \lambda_{v,k}) + \sum_{k=j+1}^{m-1} pp[v, k] / (\lambda_{v,j} - \lambda_{v,k});$$

$$s3[v\_, j\_, m\_] := pp[v, j] \left(\sum_{k=1}^{j-1} dd[v, j, k, m] + \sum_{k=j+1}^{m} dd[v, j, k, m] \right);$$

$$s4[v\_, j\_] := pp[v, j] \times u1[\lambda_{v,j}, \lambda_{v,j}];$$

$$s5[v\_, j\_, m\_] := \sum_{k=1}^{j-1} (pp[v, k] / (\lambda_{v,j} - \lambda_{v,k}) - pp[v, k] \times u1[\lambda_{v,j}, \lambda_{v,k}]) \, \varphi[\lambda_{v,k}] +$$

$$\sum_{k=j+1}^{m-1} (pp[v, k] / (\lambda_{v,j} - \lambda_{v,k}) - pp[v, k] \times u1[\lambda_{v,j}, \lambda_{v,k}]) \, \varphi[\lambda_{v,k}];$$

$$s6[v\_, j\_, m\_] :=$$

$$\left(\sum_{k=1}^{j-1} pp[v, j] \times dd[v, j, k, m] \times \varphi[\lambda_{v,k}] + \sum_{k=j+1}^{m} pp[v, j] \times dd[v, j, k, m] \times \varphi[\lambda_{v,k}] \right)$$

$$s7[v\_, j\_, n\_, m\_] := \sum_{\sigma=1}^{v-1} \sum_{k=1}^{m-1} (pp[\sigma, k] / (\lambda_{v,j} - \lambda_{\sigma,k}) - pp[\sigma, k] \times u1[\lambda_{v,j}, \lambda_{\sigma,k}]) \, \varphi[\lambda_{\sigma,k}] +$$

$$\sum_{\sigma=v+1}^{n} \sum_{k=1}^{m-1} (pp[\sigma, k] / (\lambda_{v,j} - \lambda_{\sigma,k}) - pp[\sigma, k] \times u1[\lambda_{v,j}, \lambda_{\sigma,k}]) \, \varphi[\lambda_{\sigma,k}];$$

```
Table[τ_σ, {σ, 1, n}]
```

$$\left\{ -1, -\frac{9}{11}, -\frac{7}{11}, -\frac{5}{11}, -\frac{3}{11}, -\frac{1}{11}, \frac{1}{11}, \frac{3}{11}, \frac{5}{11}, \frac{7}{11}, \frac{9}{11} \right\}$$

```
m = 4
```

4

```
Do[x_k = (k - 1) / (m - 1), {k, 1, m}]
Table[x_k, {k, 1, m}]
```

$$\left\{ 0, \frac{1}{3}, \frac{2}{3}, 1 \right\}$$

```
h = 2 / n
```

$$\frac{2}{11}$$

```
Do[λσ,k = τσ + h xk, {σ, 1, n}, {k, 1, m}]
Table[N[λσ,k], {σ, 1, n}, {k, 1, m}]
```

```
{{-1., -0.939394, -0.878788, -0.818182},
 {-0.818182, -0.757576, -0.69697, -0.636364}, {-0.636364, -0.575758, -0.515152, -0.454545},
 {-0.454545, -0.393939, -0.333333, -0.272727}, {-0.272727, -0.212121, -0.151515, -0.0909091},
 {-0.0909091, -0.030303, 0.030303, 0.0909091}, {0.0909091, 0.151515, 0.212121, 0.272727},
 {0.272727, 0.333333, 0.393939, 0.454545}, {0.454545, 0.515152, 0.575758, 0.636364},
 {0.636364, 0.69697, 0.757576, 0.818182}, {0.818182, 0.878788, 0.939394, 1.}}
```

```
ω[t_, ν_, m_] := Product[t - λν,k, {k, 1, m}]
```

```
τn+1 = 1
Do[p[σ_, k_] := (1 / (π D[ω[t, σ, m], t] /. t → λσ,k))
    Integrate[ω[t, σ, m] / (√(1 + t) / (1 - t) (t - λσ,k)), {t, τσ, τσ+1}], {σ, 1, n}, {k, 1, m}]
```

```
1
```

```
Table[N[p[σ, k]], {σ, 1, n}, {k, 1, m}]
```

```
{{0.123975, 0.163026, 0.062766, 0.0282306}, {0.0215658, 0.0616458, 0.0486247, 0.0161447},
 {0.014802, 0.0432979, 0.0370224, 0.0122269}, {0.0114846, 0.0338369, 0.0298241, 0.00984414},
 {0.00933553, 0.0275907, 0.0246399, 0.00813097}, {0.00774011, 0.0229068, 0.020535, 0.00677257},
 {0.00644773, 0.0190881, 0.0170481, 0.0056158}, {0.0053279, 0.0157635, 0.0138973, 0.00456689},
 {0.00429458, 0.0126838, 0.0108527, 0.00354756}, {0.00326642, 0.00961245, 0.00760546, 0.00244575},
 {0.00208228, 0.00620007, 0.00301113, 0.000673434}}
```

```
pp[1, 1] = N[p[1, 1]]
pp[n, m] = N[p[n, m]]
```

```
0.123975
```

```
0.000673434
```

```
Do[pp[σ, m - 1] = pp[σ, 0], {σ, 1, n - 1}]
Do[pp[σ, k] = p[σ, k], {σ, 1, n}, {k, 1, m - 1}]
Do[pp[σ, 1] = p[σ - 1, m] + p[σ, 1], {σ, 2, n - 1}]
Do[pp[σ - 1, m] = pp[σ, 1], {σ, 2, n}]
Do[pp[n, k] = N[p[n, k]], {k, 1, m}]
```

```
Table[N[pp[σ, k]], {σ, 1, n}, {k, 1, m}]
```

```
{{0.123975, 0.163026, 0.062766, 0.0497964}, {0.0497964, 0.0616458, 0.0486247, 0.0309467},
 {0.0309467, 0.0432979, 0.0370224, 0.0237115}, {0.0237115, 0.0338369, 0.0298241, 0.0191797},
 {0.0191797, 0.0275907, 0.0246399, 0.0158711}, {0.0158711, 0.0229068, 0.020535, 0.0132203},
 {0.0132203, 0.0190881, 0.0170481, 0.0109437}, {0.0109437, 0.0157635, 0.0138973, 0.00886148},
 {0.00886148, 0.0126838, 0.0108527, 0.00681398}, {0.00681398, 0.00961245, 0.00760546, 0.00208228},
 {0.00208228, 0.00620007, 0.00301113, 0.000673434}}
```

```
dd[v_, j_, k_, m_] :=
```

$$\left(\text{Delete}\left[\prod_{i=1}^{m}(\lambda_{v,j}-h_{v,i}),\ \{\{j\},\{k\}\}\right]\ /.\ h\to\lambda\right)\Big/\left(\text{Delete}\left[\prod_{i=1}^{m}(\lambda_{v,k}-h_{v,i}),\ \{\{k\}\}\right]\ /.\ h\to\lambda\right)$$

```
lag[f_, x_, m_] :=
```

$$\sum_{k=1}^{n}(\text{Product}[x-h_i,\{i,1,m\}]\ f[h_k])\ /\ ((x-h_k)\ (D[\text{Product}[x-h_i,\{i,1,m\}],x]\ /.\ x\to h_k))$$

```
Do[Print[N[((1 + s1[v, j, n, m] + s2[v, j, m] - s3[v, j, m] + s4[v, j]) φ[λ_v,j] -
    s5[v, j, m] + s6[v, j, m] - s7[v, j, n, m]] == 2,], {v, 1, n - 1}, {j, 1, m - 1}]

Do[Print[N[((1 + s1[n, j, n, m] + s2[n, j, m] - s3[n, j, m] + s4[n, j]) φ[λ_n,j] - s5[n, j, m] + s6[n, j, m] -
    s7[n, j, n, m]] == 2,], {j, 1, m}]
```

-7.32028 φ[-1.] + 8.82737 φ[-0.939394] - 2.55003 φ[-0.878788] + 0.956357 φ[-0.818182] +
  0.255291 φ[-0.757576] + 0.161436 φ[-0.69697] + 0.085837 φ[-0.636364] + 0.103239 φ[-0.575758] +
  0.0774949 φ[-0.515152] + 0.0442775 φ[-0.454545] + 0.0570901 φ[-0.393939] +
  0.0459382 φ[-0.333333] + 0.0272022 φ[-0.272727] + 0.0362923 φ[-0.212121] +
  0.030245 φ[-0.151515] + 0.0182765 φ[-0.0909091] + 0.024862 φ[-0.030303] + 0.0210921 φ[0.030303] -
  0.012897 φ[0.0909091] + 0.0177431 φ[0.151515] + 0.0151431 φ[0.212121] + 0.00931325 φ[0.272727] +
  0.0128827 φ[0.333333] + 0.0109303 φ[0.393939] + 0.00672045 φ[0.454545] + 0.00929189 φ[0.515152] +
  0.00769238 φ[0.575758] + 0.00468 φ[0.636364] + 0.00640619 φ[0.69697] + 0.00492455 φ[0.757576] +
  0.0013115 φ[0.818182] + 0.00380267 φ[0.878788] + 0.00180023 φ[0.939394] == 2Null

-2.9417 φ[-1.] - 1.03587 φ[-0.939394] + 3.72635 φ[-0.878788] - 0.0366922 φ[-0.818182] +
  0.340287 φ[-0.757576] + 0.201729 φ[-0.69697] + 0.102968 φ[-0.636364] + 0.120398 φ[-0.575758] +
  0.0885261 φ[-0.515152] + 0.0497874 φ[-0.454545] + 0.0633981 φ[-0.393939] +
  0.0505006 φ[-0.333333] + 0.0296547 φ[-0.272727] + 0.0392866 φ[-0.212121] +
  0.0325441 φ[-0.151515] + 0.0195639 φ[-0.0909091] + 0.0264927 φ[-0.030303] +
  0.0223858 φ[0.030303] + 0.0136394 φ[0.0909091] + 0.0187049 φ[0.151515] + 0.0159182 φ[0.212121] +
  0.00976455 φ[0.272727] + 0.0134751 φ[0.333333] + 0.0114081 φ[0.393939] +
  0.0070003 φ[0.454545] + 0.00966108 φ[0.515152] + 0.00798444 φ[0.575758] +
  0.00485003 φ[0.636364] + 0.0066292 φ[0.69697] + 0.00508899 φ[0.757576] +
  0.00135356 φ[0.818182] + 0.00391989 φ[0.878788] + 0.00185363 φ[0.939394] == 2Null

-0.84915 φ[-1.] - 3.72356 φ[-0.939394] + 2.77848 φ[-0.878788] + 1.16785 φ[-0.818182] +
  0.510039 φ[-0.757576] + 0.268761 φ[-0.69697] + 0.128605 φ[-0.636364] + 0.144356 φ[-0.575758] +
  0.103189 φ[-0.515152] + 0.0568472 φ[-0.454545] + 0.0712532 φ[-0.393939] + 0.0560539 φ[-0.333333] +
  0.0325845 φ[-0.272727] + 0.0428085 φ[-0.212121] + 0.0352129 φ[-0.151515] +
  0.0210414 φ[-0.0909091] + 0.0283461 φ[-0.030303] + 0.0238436 φ[0.030303] +
  0.0144697 φ[0.0909091] + 0.0197732 φ[0.151515] + 0.016774 φ[0.212121] + 0.0102601 φ[0.272727] +
  0.0141223 φ[0.333333] + 0.0119279 φ[0.393939] + 0.0073035 φ[0.454545] + 0.0100596 φ[0.515152] +
  0.00829861 φ[0.575758] + 0.00503237 φ[0.636364] + 0.00686764 φ[0.69697] + 0.00526435 φ[0.757576] +
  0.00139829 φ[0.818182] + 0.0040443 φ[0.878788] + 0.00191017 φ[0.939394] == 2Null

{{φ[-1.] → 2.07293, φ[-0.939394] → 2.06444, φ[-0.878788] → 2.05616, φ[-0.818182] → 2.0481,
  φ[-0.757576] → 2.04025, φ[-0.69697] → 2.03262, φ[-0.636364] → 2.0252, φ[-0.575758] → 2.018,
  φ[-0.515152] → 2.01101, φ[-0.454545] → 2.00424, φ[-0.393939] → 1.99767,
  φ[-0.333333] → 1.99131, φ[-0.272727] → 1.98516, φ[-0.212121] → 1.97921,
  φ[-0.151515] → 1.97346, φ[-0.0909091] → 1.96791, φ[-0.030303] → 1.96255, φ[0.030303] → 1.95738,
  φ[0.0909091] → 1.95239, φ[0.151515] → 1.94759, φ[0.212121] → 1.94296, φ[0.272727] → 1.93851,
  φ[0.333333] → 1.93423, φ[0.393939] → 1.93011, φ[0.454545] → 1.92616, φ[0.515152] → 1.92236,
  φ[0.575758] → 1.91871, φ[0.636364] → 1.91521, φ[0.69697] → 1.91185, φ[0.757576] → 1.90863,
  φ[0.818182] → 1.90555, φ[0.878788] → 1.90259, φ[0.939394] → 1.89977, φ[1.] → 1.89701}}

References

1. Abramov, B. D., Matveev, A. F. On the Reduction of Boundary Problems in Neutron Transport Theory to Singular Integral Equations. Preprint (ITEP), Moscow, 1987, No. 46, 32 pp.

2. Babkin, V. I., Belotserkovsky, S. M., Gulyaev, V. V., Dvorak, A. V. Flows and Bearing Surfaces. Computer Simulation. Moscow: Nauka, 1989, 208 pp.

3. Bantsuri, R. D. Solution of the First Fundamental Problem and a Mixed Problem of Elasticity Theory for an Infinite Strip Cut Along a Semiline. // Reports of the Academy of Sciences of the Georgian SSR, 1965, Vol. 37, No. 2, pp. 275-281.

4. Bantsuri, R. D. Solution of the First Fundamental Problem of Elasticity Theory for a Wedge with a Finite Crack. // Doklady of the Academy of Sciences of the USSR, 1966, Vol. 167, No. 6, pp. 1256-1259.

5. Barenblatt, G. I., Cherepanov, G. P. On Brittle Cracks in Longitudinal Shear. // PMM, 1961, Vol. 25, No. 6, pp. 1110-1119.

6. Barenblatt, G. I., Cherepanov, G. P. On the Influence of Body Boundaries on the Development of Cracks in Brittle Failure. // Izv. Acad. Sci. USSR, Mechanics and Machine Engineering, 1960, Vol. 3, pp. 79-88. Note to this work. // Izv. Acad. Sci. USSR, Mechanics and Machine Engineering, 1962, Vol. 1, p. 153.

7. Belokopytova L. V., Fil'shtinskiy L. A., Two-Dimensional Boundary Problem of Electroelasticity for a Piezoelectric Medium with Cracks.PMM, 1979, Vol. 43, No. 1, pp. 138-143.

8. Belozertcovsky S. M., Horseshoe Vortex in Unsteady Motion. PMM, 1955, Vol. XIX, Issue 4, pp. 410-420.

9. Belozertcovsky S. M., Spatial Unsteady Motion of Bearing Surfaces. PMM, 1955, Vol. XIX, Issue 4, pp. 410-420.

10. Belozertcovsky S. M., Research on the Aerodynamics of Modern Bearing Surfaces.Dissertation for the Doctor of Technical Sciences. Moscow: 1955. 1955

11. Belozertcovsky S. M., Thin Bearing Surface in Subsonic Gas Flow. Moscow: Nauka, 1965, 244pp.

12. Belozertcovsky S. M., Calculation of Flow Around Wings of Arbitrary Planform Shape in a Wide Range of Angles of Attack. Izv. Acad. Sci. USSR, MGG, 1968, No. 4.

13. Belozertcovsky S. M., Skripach B. K., Aerodynamic Derivatives of an Aircraft and Wing at Subsonic Speeds. Moscow: Nauka, 1975, 424 pp.

14. Belozertcovsky S. M., Nisht M. I., Separated and Non-Separated Flow Around Thin Wings in an Ideal Fluid. Moscow: Nauka, 1978, 352 pp.

15. Belozertcovsky O. M., Belozertcovsky S. M., Davydov Y. M., Nisht M. I., Simulation of Separated Flows on a Computer. Moscow: 1984, 122 pp.

16. Belozertcovsky S. M., Lifanov I. K., Numerical Methods in Singular Integral Equations. Moscow: Nauka, 1985, 256 pp.

17. Belozertcovsky S. M., Nisht M. I., Ponomarev A. T., Rysev O. V., Study of Parachutes and Deltoid Planes on a Computer. Moscow: Mashinostroenie, 1987, 240 pp.

18. Belozertcovsky S. M., Kotovsky V. N., Nisht M. I., Fedorov R. M., Mathematical Modeling of Planar-Paralel Separated Flow Around Bodies. Moscow: Nauka, 1988, 232 pp.

19. Berezhnitsky L. T., On the Limit Equilibrium of a Plate Weakened with a System of Cracks Located Along a Straight Line at an Angle in the Direction of Tension. In: Concentration of Stresses, Volume 1. Naukova Dumka, Kyiv, 1965, pp. 46-52.

20. Boykov I. V., Passive and Adaptive Algorithms for Approximate Computation of Singular Integrals, Part 1-2. Penza State Technical University Press, Penza, 1995.

21. Brown W., Crawley J., Testing of High-Strength Metallic Materials for Viscosity Fracture in Plane Deformation. Mir, Moscow, 1972.

22. Gabdulkhayev B. G., Optimal Approximations of Solutions to Linear Problems. Kazan: Kazan University Press, 1980, 232 pp.

23. Gandel Y. V., Lifanov I. K., Matveev A. F., Numerical Solution of Mixed Boundary Problems in Mathematical Physics Reducing to Singular Integral Equations on a System of Segments.Preprint ITEF, No. 174, Moscow, 1984, 55 pp.

24. Gandel Y. V., Lifanov I. K., On the Application of the Ideas of the Discrete Vortex Method to Electrodynamics Problems. Scientific and Methodological Materials on Numerical Methods. Moscow, Zhukovsky Air Force Engineering Academy, 1985, pp. 3-13.

25. Gandel Y. V., Lifanov I. K., On the Solution of Singular Integral Equations in the Robin Problem. TFFA and Their Applications, Kharkiv, Vyscha Shkola, 1986, Issue 46, pp. 18-21.

26. Gakhov F. D., Boundary Problems. Moscow: Publishing House of Physical and Mathematical Literature, 1963, 639 pp.

27. Golubev V. V., Lectures on Wing Theory. Moscow - Leningrad: GITTL, 1949, 480 pp.

28. Djishkiani A. V., On the Solution of Singular Integral Equations by Approximate Projection Methods. Journal of Applied Mathematics and Mathematical Physics, 1979, Vol. 19, No. 5, pp. 1149-1161.

29. Djishkiani A. V., On the Solution of Singular Integral Equations by Collocation Methods.Journal of Applied Mathematics and Mathematical Physics, 1981, Vol. 21, No. 2, pp. 355-362.

30. Djishkiani A. V., On the Approximate Solution of a Class of Singular Integral Equations.Proceedings of the Tbilisi Mathematical Institute, 1986, Vol. 86, pp. 41-49.

31. Dmitriev V. I., Zakharov E. V., Integral Equations in Boundary Problems of Electrodynamics. Moscow: MSU Press, 1987, 168 pp.

32. Duduchava R. V., Shargorodsky E. M., On Some Singular Integral Operators with Fixed Singularities. Proceedings of the Tbilisi Mathematical Institute named after A. M. Razmadze, Vol. 93, Boundary Problems of the Theory of Functions with Compressible Variables and Special Integral Equations, 1, 1990, pp. 1-32.

33. Dushkov P. N., On Direct Methods for Solving Singular Integral Equations of the First Kind.Izv. Vuzov. Matematika, 1973, No. 7, pp. 12-24.

34. Ekobori T., Physics and Mechanics of Fracture and Strength of Solids.Metallurgiya, Moscow, 1971.

35. Zakharov E. V., Pimenov Y. V., Numerical Analysis of Radio Wave Diffraction.Moscow: Radio i Svyaz, 1982, 184 pp.

36. Ivanov V. V., Theory of Approximate Methods and Its Application to the Numerical Solution of Singular Integral Equations. Kyiv: Naukova Dumka, 1968, 286 pp.

37. Kalandia A. I., Mathematical Methods in Two-Dimensional Elasticity. Moscow: Nauka, 1973, 303 pp.

38. Kantorovich L. V., Akilov G. P., Functional Analysis. Moscow: Nauka, 1977, 744 pp.

39. Kapanadze, G. On a Mixed Problem of the Theory of Analytic Functions for a Circular Ring Cut Along Circular Arcs. //GTU, Proceedings of the International Symposium on Problems of Thin-Walled Spatial Systems. July 4-5, 2001, Tbilisi, pp. 17-21.

40. Kveselava D. A., Samsonia Z. V., On Conformal Moduli of Close Doubly Connected Domains. Proceedings of the VTs of the Academy of Sciences of the Georgian SSR, Vol. 1, 1960, pp. 5-13.

41. Kolosov G. V., On an Application of the Theory of Complex Functions to the Planar Problem of Mathematical Elasticity Theory. Typ. K. Mattisev, Yuriev, 1909.

42. Colton D., Kress R., Methods of Integral Equations in Scattering Theory. Moscow: Mir, 1987, 311 pp.

43. Korneichuk A. A., Quadrature Formulas for Singular Integrals. In: Numerical Methods for Solving Differential and Integral Equations and Quadrature Formulas. Moscow: Nauka, 1964, pp. 64-74.

44. Krylov V. I., Approximate Calculation of Integrals. Moscow: Nauka, 1967, 500 pp.

45. Kublashvili M. D., On the Stability of Some Quadrature Formulas for Singular Integrals.All-Union Symposium: The Method of Discrete Singularities in Mathematical Physics Problems, Abstracts of Reports, Kharkiv, 1985, p. 55.

46. Kublashvili M. D., A Comment on One Scheme for the Numerical Solution of Singular Integral Equations with Open Contours of Integration. Proceedings of the IVM of the Academy of Sciences of the Georgian SSR, Vol. XXV:1, 1985, pp. 67-73.

47. Kublashvili M. D., On Uniform Estimates of Approximation of Singular Integrals with Open Contours by Piecewise-Interpolating Functions. Scientific Works of the Georgian Polytechnic Institute named after V. I. Lenin. Mathematical Analysis, Tbilisi, 1989, pp. 98-106.

48. Kublashvili, M. On the Numerical Solution of Singular Integral Equations of the First Kind. // Proceedings of the First Republican Scientific Conference of Professors, Teachers, and Students. Abastumani, May 18-20, 1995, p. 29.

49. Kublashvili, M. On the Numerical Solution of the Crack Problem. // Proceedings of the International Scientific Conference "Automated Control Systems". Tbilisi, September 27-28, 1996, pp. 28-31.

50. Kublashvili, M. D. Approximate Calculation of Singular Integrals with Piecewise-Differentiable Densities. // Proceedings of the International Symposium Dedicated to the Problems of Thin-Walled Spatial Systems. Tbilisi, July 4-5, 2001, pp. 125-131.

51. Kublashvili, M. D. On the Numerical Solution of Some Problems of Infinite Plates with Cracks. // Proceedings of the International Symposium Dedicated to the Problems of Thin-Walled Spatial Systems. Tbilisi, July 4-5, 2001, pp. 132-136.

52. Kublashvili, M. D. On the Numerical Solution of the Problem of a Thermally Insulated Crack. // International Scientific Journal "Problems of Applied Mechanics". No. 4(9), 2002, pp. 89-91.

53. Kublashvili, M. Approximate Calculation of Cauchy-Type Integrals for Open Contours Using a Correctable Parameter and Some of Their Properties. // Science and Technologies, Tbilisi, No. 10-12, 2002, pp. 54-58.

54. M. Kublashvili, On the Numerical Solutions of the Singular Integral Equation for the Problem of a Cylindrical Shell under the Influence of a Tangential Load. In: Proceedings of the Open Scientific-Technical Conference Dedicated to the 80th Anniversary of the Professors and Lecturers of the STU, Tbilisi, 2002, p. 20.

55. M. Kublashvili, Numerical Solution of the First Kind Singular Integral Equations with Discrete Singularities Method for Open Contours. In: Proceedings of the Open Scientific-Technical Conference Dedicated to the 80th Anniversary of the Professors and Lecturers of the STU, Tbilisi, 2002, p. 20.

56. M. Kublashvili, Approximation of Cauchy Type Integrals for Open Contours and Some of Their Applications. In: Proceedings of the Open Scientific-Technical Conference Dedicated to the 80th Anniversary of the Professors and Lecturers of the STU, Tbilisi, 2002, p. 214.

57. Kublashvili M. D., Numerical Solution of the Thin Airfoil Problem Using a Singular Integral Equation. In: International Scientific Journal "Problems of Applied Mechanics", No. 2(11), 2003, pp. 119-122.

58. Kublashvili M. D., On the Numerical Solution of a Class of Singular Integral Equations. In: International Scientific Journal GEN, No. 1, 2003, pp. 38-40.

59. Kublashvili M. D., Numerical Solution of the Longitudinal Shear Crack Problem in an Elastic Body. In: International Scientific Journal GEN, No. 1, 2003, pp. 41-43.

60. Kublashvili M. D., On the Construction of a Higher-Order Accuracy Computational Scheme of the Discrete Vortex Method in the Case of Open Contours. In: International Scientific Journal GEN, No. 1, 2003, pp. 44-46.

61. Kupradze V. D., Boundary Problems of the Theory of Vibrations and Integral Equations. Leningrad.

62. Kutateladze G. A., Some Boundary Problems of the Theory of Analytic Functions for a Circle with Parazrezon. In: Proceedings of the Institute of Applied Mathematics (TGU) Named after I. N. Vekua, Vol. 4, No. 1, 1990, pp. 110-113.

63. Kutateladze G. A., On the Torsion of a Circular Prismatic Beam with Longitudinal Cuts. In: Symposium on Problems of Mechanics of Continuous Media, Tbilisi, TGU, 1972, pp. 188-193.

64. Lavrentiev M. A., On the Construction of a Flow Around a Curve of Given Shape. Proceedings of TsAGI, Vol. 118, 1932, pp. 1-56. GITTL, 1950, 280.

65. Lifanov I. K., Polonsky Y. E., Justification of the Numerical Method of Discrete Vortices for Solving Singular Integral Equations. PMM, 1975, Vol. 39, No. 4, pp. 742-746.

66. Lifanov I. K., The Method of Singular Integral Equations and Numerical Experiment. M: TOO "Janus", 1995, 504 pages.

67. Lifanov I. K., On Singular Integral Equations with One-Dimensional and Multiple Cauchy-Type Integrals. DAN USSR, 1978, Vol. 239, No. 2, pp. 265-268.

68. Lifanov I. K., On the Method of Discrete Vortices. Applied Mathematics and Mechanics, 1979, Vol. 43, No. 1, pp. 184-188.

69. McClintock F., Argon A., Deformation and Fracture of Materials. Mir, Moscow, 1970.

70. Mirianashvili M. G., On the Principle of Close Domains and Its Applications.Proceedings of the Institute of Mathematics, Academy of Sciences of the Georgian SSR, Vol. XXIX: 1, 1985, pp. 169-178.

71. Mikhin S. G., Integral Equations. Moscow-Leningrad, OGIIZ, State Publishing House of Technical-Theoretical Literature, 1949, 378 pages.

72. Morozov N. F., Mathematical Issues of the Theory of Cracks. Moscow, NAUKA, 1984, 255 pages.

73. Musayev B. I., Approximate Solution of the Full Singular Integral Equation on a Segment.Institute of Cybernetics, Academy of Sciences of the Azerbaijani SSR, Baku, 1985, 34 pages, Dep. at VINITI, 23.10.85, No. 73, pp. 77-85.

74. Musayev B. I., On the Approximate Solution of Singular Equations. Preprint, Institute of Physics, Academy of Sciences of the Azerbaijani SSR, Baku, 1986, No. 17, 48 pages.

75. Musayev B. I., On the Approximate Solution of Singular Integral Equations. In: Singular Integral Operators, Azerbaijan State University, Baku, 1986, pp. 33-61.

76. Musayev B. I., Approximate Solution of Singular Integral Equations for Negative Index Using the Method of Mechanical Quadratures. Reports of the Academy of Sciences of the USSR, 1988, Vol. 298, No. 2, pp. 286-290.

77. Muskhelishvili N. I., Singular Integral Equations. Moscow: Nauka, 1968, 511 pages.

78. Muskhelishvili N. I., Some Basic Problems of the Mathematical Theory of Elasticity. Moscow: Nauka, 1966, 707 pages.

79. Nazarchuk Z. T., Numerical Study of Wave Diffraction on Cylindrical Structures. Kiev: Naoukova Dumka Publishing House, 1989, 256 pages.

80. Panasuk V. V., Berezhnitskiy L. T., Kovchik S. E., On the Development of Arbitrarily Oriented Straight Crack on a Plate in Tension. PM, 1965, Vol. 1, Issue 2, pp. 48-55.

81. Panasuk V. V., Ultimate Equilibrium of Brittle Bodies with Cracks. Naoukova Dumka, Kiev, 1968.

82. Panasuk V. V., Savryuk M. P., Datsyshyn A. P., Stress Distribution Near Cracks in Plates and Shells. Naoukova Dumka, Kiev, 1976, 443 pages.

83. Panasuk V. V., Savryuk M. P., Nazarchuk Z. T., Method of Singular Integral Equations in Two-Dimensional Diffraction Problems. Kiev: Naoukova Dumka, 1984, 344 pages.

84. Parton V. Z., Morozov E. M., Mechanics of Elastic-Plastic Fracture. Naouka, Moscow, 1974.

85. Parton V. Z., Perlin P. I., Integral Equations in the Theory of Elasticity. Naouka, Moscow, 1977, 312 pages.

86. Parton V. Z., Perlin P. I., Methods of Mathematical Theory of Elasticity. Naouka, Moscow, 1981, 688 pages.

87. Polonskiy Ya. E., Justification of the Numerical Method "Discrete Vortices" for Solving Singular Integral Equations. Prikl. Mat. i Mekh., 1975, Vol. 39, No. 4, pp. 742-746.

88. Pykhteev G. N., On the Computation of Some Singular Integrals with Cauchy-Type Kernel.Prikl. Mat. i Mekh., 1959, Vol. 23, Issue 6, pp. 1078-1082.

89. Pykhteev G. N., On the Computation of Some Integrals with a Regular Cauchy-Type Kernel.Prikl. Mat. i Mekh., 1960, Vol. 24, Issue 6, pp. 1114-1122.

90. Pykhteev G. N., On the Quadrature Formulas for Cauchy-Type Integrals along a Straight Open Contour and Methods for Estimating Their Errors. Izv. AN BSSR, Ser. Phys.-Mat. Sci., 1969, No. 5, pp. 55-63.

91. Pykhteev G. N., On the Construction of Quadrature Formulas for Cauchy-Type Integrals along a Straight Open Contour. Izv. AN BSSR, Ser. Phys.-Mat. Sci., 1970, No. 2, pp. 61-69.

92. Savruk M. P., Two-Dimensional Elasticity Problems for Bodies with Cracks. Naouka, Kyiv, 1981, 324 pages.

93. Sanikidze D. G., On the Approximate Computation of Curvilinear Singular Integrals. Annotations of the Seminar of the Institute of Applied Mathematics, Tbilisi State University, No. 3, 1970, pp. 15-17.

94. Sanikidze D. G., Approximate Solution of Singular Integral Equations for Closed Contours of Integration. Annotations of the Seminar of the Institute of Applied Mathematics, Tbilisi State University, No. 5, 1971, pp. 31-35.

95. Sanikidze D. G., On the Order of Approximation of Singular Integrals for a Certain Class of Densities by Sums of Complicated Type. Reports of the Academy of Sciences of the Georgian SSR, 68, No. 3, 1972, pp. 533-536.

96. Sanikidze D. G., On the Numerical Solution of Boundary Problems Using the Approximation of Singular Integrals. Differential Equations, 1993, Vol. 29, No. 9, pp. 1632-1644.

97. Sanikidze D. G., On the Justification of the Boundary Integral Equation Method for Domains with Nonsmooth Boundaries. Differential Equations, 1996, Vol. 32, No. 9, pp. 1153-1159. Appl. Math. and Informatics, Vol. 1, No. 1, 1996, pp. 148-152.

98. Sanikidze D. G., Mirianashvili M. G., On the Application of the Discrete Vortex Method to the Numerical Solution of Singular Integral Equations with Smooth Closed Contours. Scientific Proceedings of the Applied Mathematics and

Mathematical Modeling Department of the Zhukovsky Air Force Engineering Academy, 1997, pp. 178-180.

99. Sanikidze D. G., On the Higher-Precision Discrete Vortex Method for the Numerical Solution of a Class of Singular Integral Equations. Differential Equations, 1998, Vol. 34, No. 9.

100. Sanikidze D. G., On Some Estimates in the Modified Scheme of the Discrete Vortex Method. Proceedings of the VII International Symposium "Methods of Discrete Singularities in Mathematical Physics Problems", Ukraine, 1999, pp. 118-120.

101. Sanikidze D. G., Khubetsi Sh. S., On the Application of External Nodes in Modified Schemes of the Discrete Vortex Method. Proceedings of the IX International Symposium "Methods of Discrete Singularities in Mathematical Physics Problems" (MDSMP-2000), Orel, 2000, pp. 395-398.

102. Sanikidze D. G., Ninidze K. R., The Method of Free Parameters in the Approximate Calculation of Cauchy-Type Integrals. Proceedings of the X International Symposium "Methods of Discrete Singularities in Mathematical Physics Problems" (MDSMP-2001), pp. 299-302.

103. Saren V. É., On the convergence of the discrete vortex method. Siberian Mathematical Journal, 1978, Vol. 19, No. 2, 385–395.

104. Sedov L. I., Mechanics of Continuous Media. Vol. 2. Nauka, Moscow, 1973.

105. Sih G. C., On the singular nature of thermal stresses at the tip of a crack. Applied Mechanics (translated proceedings of the American Society of Mechanical Engineers), Izd. Lit., 1962, Vol. 29E, No. 3, 157–159.

106. Stark I., Generalized quadrature formula for Cauchy integrals. Rocket Technology and Cosmonautics, 1971, No. 9, 244–245.

107. Cherepanov G. P., A problem on the indentation of an indenter with crack formation. Applied Mathematics and Mechanics (PMM), 1963, Vol. 27, No. 1, 150–153.

108. Cherepanov G. P., Mechanics of Brittle Fracture. Nauka, Moscow, 1974.

109. Sheshko M. A., On the numerical solution of singular integral equations of the first kind. Differential Equations, 1977, Vol. 13, No. 8, 1493–1502.

110. R. Bansturi. One one mixed type bounday value problem of the theory of analitic functions. Proceedings of A. Razmadze matema-tical institute. Vol. 121, 3-9, 1999.

111. Griffith A. A., The phenomenon of rupture and flow in solids. -Phil. Trans. Roy. Soc., 1920, A 221, 163-198.

112. Griffith A. A., The theory of rupture. -proc. First Intern. Congr. Appl. Mech., Delft, 1924, 55-63.

113. Dang Dac Q., Nurrie D. H. A finite element method for the solution of singular integral equations, Comput. And Math. 1978, 4, №3, 219-224.

114. England A. H., A note on cracks under longitudinal shear. Matematika, 1963, 10, 20, 107-11.

115. Erdogan F. E., Gupa G. D. On the numerical solution of singular in-tegral equation. // Quart. Apll. Math. 1972, V29, 525-534.

116. Erdogan F. E., Gupta G. D., Cook T.S. The numerical solutions of singular integral equations. -In: Methods of Analysis and Solutions of Crack Problems. Noordhoff Intern. Publ., Leyden, 1973, 368-425.

117. Irwin G. R., Wells A. A., A continuum mechanics view of crack propagation. // Metallurg. Revs, 1965, 10, 38, 223-270.

118. Krenk S. Polunomial Solutions to Singular Integral Equations// Riso National Laboratory, DK-4000 Rosialde, Denmark, January 1981., 1269-1275.

119. S. Krenk Polynivial Solytion to Singular Integral Equations, with Applications to Elasticity Teory, Technical University of Denmark, Thesis of disertation for the Degree of Doctor Tecnices, 1981.